AF567021

Fundamentals of Engineering High-Performance Actuator Systems

Other SAE books of interest:

Advances in Aircraft Landing Gear
Robert Kyle Schmidt
(Product Code: PT-169)

Advances in Aircraft Brakes and Tires
Robert Kyle Schmidt
(Product Code: PT-171)

Advanced Engine Development at Pratt & Whitney
Dick Mulready
(Product Code: R-252)

For more information or to order a book, contact:
SAE International
400 Commonwealth Drive
Warrendale, PA 15096, USA

Phone: 1+877.606.7323 (U.S. and Canada only)
or 1+724.776.4970 (outside U.S. and Canada)
Fax: 1+724.776.0790
Email: CustomerService@sae.org
Website: books.sae.org

Fundamentals of Engineering High-Performance Actuator Systems

By Kenneth W. Hummel

Warrendale, Pennsylvania, USA

400 Commonwealth Drive
Warrendale, PA 15096 USA
E-mail: CustomerService@sae.org
Phone: +1 877.606.7323 (inside USA and Canada)
+1 724.776.4970 (outside USA)
Fax: +1 724.776.0790

Copyright © 2017 SAE International. All rights reserved.

No part of this publication may be reproduced, stored in a retrieval system, distributed, or transmitted, in any form or by any means without the prior written permission of SAE International. For permission and licensing requests, contact SAE Permissions, 400 Commonwealth Drive, Warrendale, PA 15096-0001 USA; email: copyright@sae.org; phone: 1+724-772-4028; fax: 1+724-772-9765.

Library of Congress Catalog Number: 2016947448
SAE Order Number R-459
http://dx.doi.org/10.4271/r-459

Information contained in this work has been obtained by SAE International from sources believed to be reliable. However, neither SAE International nor its authors guarantee the accuracy or completeness of any information published herein and neither SAE International nor its authors shall be responsible for any errors, omissions, or damages arising out of use of this information. This work is published with the understanding that SAE International and its authors are supplying information, but are not attempting to render engineering or other professional services. If such services are required, the assistance of an appropriate professional should be sought.

ISBN-Print 978-0-7680-8362-0
ISBN-PDF 978-0-7680-8363-7
ISBN-epub 978-0-7680-8365-1
ISBN-prc 978-0-7680-8364-4

To purchase bulk quantities, please contact SAE Customer Service:

Email: CustomerService@sae.org
Phone: 1+877-606-7323 *(inside USA and Canada)*
1+724-776-4970 *(outside USA)*
Fax: 1+724-776-0790

Visit the SAE International Bookstore at books.sae.org

Contents

Chapter 1
Introduction

Today, actuators play an increasingly important role in our society. They are the elements that make everything around us move, increasingly by remote control through software commands. Robotic devices can have multiple actuators for independent motion in many degrees of freedom. Less sophisticated systems may only have a single actuator to perform the motion required. In all cases, the actuator needs to be engineered correctly to perform its function reliably and at the lowest achievable cost.

Actuators are the key to allowing machines to become more sophisticated and perform complex tasks that were previously done by humans. Now a multitude of tasks, large and small, are performed by machines. Some tasks are performed with complete automation where the operator just designates the task to be done and the machine completes it without any further input from the operator. Other tasks are done with various levels of input from the operator to guide the machine and make decisions along the way. In either case, actuators need to provide motion in a safe, controlled manner.

The fundamental approach to actuator design is the same for large or small loads. It is really just a matter of scaling. Of course, the engineer needs to utilize technologies relevant for the size of the task, but as technologies mature, an approach that was not feasible a few years ago may now be the best choice. For example, a few years ago, the best actuator for moving large loads were hydraulic based, but now electric drive options are available and may be the best choice for many applications. Most readily available commercial actuators are for mid-size applications. Very large or very small applications require additional design effort to create custom components to satisfy the design requirements. As the designer gains experience with the design process, components available, and manufacturing/assembly constraints, the ability to transfer this knowledge to other large and small designs increases.

As defined in this book, actuator design is a subset of mechanical design. It involves engineering the mechanical components necessary to make a product move as desired. This includes defining loads, move profiles, associated mechanisms, and prime movers.

In addition, operating and storage environments need to be defined and considered so that the end product has satisfactory behavior and reliability. All these should be defined in a stable set of requirements that can be used to guide the design and judge its maturity.

The main difference between a designed product and an engineered product is the level of knowledge about how the product is going to perform before it is built. A designed product may look good and can perform as the user/customer expects, but it may be bigger or heavier or use more energy than necessary because its design is based on limited analysis. On the other hand, an engineered product's design and features are based on engineering analysis, so performance, weight, and energy usage are optimized. From the engineer's viewpoint, customers include internal company representatives from management, sales, and marketing. External customers include the person or organization buying the actuator as well as users who may not directly buy the machine or equipment but work for the purchaser and can influence the purchaser's decisions.

This book is written as a text to supplement actuator design courses and a reference to engineers involved in the design of high-performance actuator systems. It highlights the design approach and features that should be considered when moving a payload at precision levels and/or speeds that are not as important in low-performance applications. However, there are specialized areas such as military and aerospace applications that may have their own set of additional or unique features or processes not addressed in this book.

Some sections of this book contain considerable equation background and derivation. The intent is to help the reader understand the basic principles so that they can be applied correctly and how equations are related. There is also a desire to minimize the mystery on how an equation is derived from the previous one.

1.1 Fundamentals

The fundamentals for actuator design include many of the basic machine design practices used in other mechanical engineering design tasks. One has to consider basics such as performance requirements, loading, constraints, component strength, stiffness, and safety margin along with the environment that the product is exposed to and must operate in. Each of these areas is briefly explored and reviewed in this section.

1.2 Performance

In the context of this book, an actuator or product's performance involves how fast it can move a payload between defined positions with the desired level of control. High performance does not necessarily mean that the payload is large or the motion is exceptionally fast, but rather that the load is consistently moved as desired with the appropriate combination of product size, mass, reliability, maintainability, efficiency, and cost. Most design efforts start with defining an approach and/or configuration that can meet the desired payload, move time or velocity, and range of motion. These parameters define configuration, forces and component size options. These options can then be compared to determine the best combination that satisfies all the other design requirements. An actuator rapidly moving a load from position A and slamming it into a stop at position B would not be defined as a high-performance actuator. An actuator that consistently moves the payload from A to B within the specified boundary conditions, over the specified range of environmental conditions (temperature, rain, snow, ice, etc.), externally imposed disturbances (mobility loads, shock, vibration), and system drive variations (hydraulic pressure, voltage variation, friction, etc.) is a high-performance actuator.

1.3 Loads

Loads that the actuator and its surrounding structure must accommodate include static loads, dynamic (transient) loads, inertial loads, and environmental loads. These loads cause the structural components to be in tension or compression or have internal moments and can be applied as concentrated (point) or distributed loads. For most engineers, static loads are the easiest ones to define and in some cases provide enough information to perform initial component sizing. Static loads may come from the structural mass, externally applied loads, and loads from any item being moved by the actuator. Dynamic loads can be generated by the actuator moving its load or can come from external influences such as movement of nearby equipment or movement of the structure that mounts the actuator. Some of these loads may be vibration loads that can excite system frequencies and need to be evaluated for high-cycle fatigue or shock loads that cause less-frequent high loads that need to be evaluated for low-cycle fatigue and strength. Inertial loads come from accelerating or decelerating the load, structure, and actuator components. One also needs to consider loads imposed upon the components when stopping due to an internally or externally declared emergency stop or stop brought about due to some system failure. In all cases, it is desired to bring the load to a carefully controlled stop. Unless the motion is slow, inertial loads need to be included in both structural and drive load calculations. Finally, environmental loads can be static loads from wind, rain, snow, or ice directly loading the components or can be thermal loads that cause the equipment to expand or contract. Some environmental loads such as wind can be transient, and as such, their dynamic effect on actuator force and structural excitation needs to be considered. Design loads are explored in more depth in chapter 3.

1.4 Constraints

Loads imposed on the actuator must be transferred through the actuator to the structure supporting the actuator. To accurately assess actuator loads and ensure it performs as expected, the designer must understand how all the loads are transferred through the entire product. This allows the designer to accurately predict actuator loads and how the interfacing structures transfer loads to and from the actuator. The type of constraint also determines the type of load the actuator needs to accommodate.

The constraints found most often in mechanical structures and mechanism are pinned, pinned with friction, fixed, and simply supported. For structural components, including actuator, both ends do not need to have the same type of constraint, but all the forces and moments need to be considered. In most cases, a pinned joint is the best connection between an actuator and its interfacing structural components. A true frictionless, pinned connection cannot transmit moments, so the actuator only sees tensile or compressive loads. While no joint is truly frictionless, they are easy to analyze and can represent an analytical boundary that can be useful. Any friction present in a pinned joint causes moment to be transmitted through the actuator which must be accommodated by the design, both structurally and functionally. A fixed joint is a common but not always desirable joint configuration found in actuators. Its behavior is similar to a pinned joint with friction except any moment or moment inducing load applied at the joint is passed through the mechanical component or actuator. All these loads must be accommodated by the design so that it operates intended over its design life. While some constraints within a product may be of the simply supported type, it is rare that an actuator has such a constraint because a simple support can only restrict movement in one direction. This usually limits the load to only being compression, though tension-only configurations are possible. Any other load on this type of boundary condition can cause unintended motion, which could result in failure.

1.5 Design Margin

The design margin applied to components of any given design compensates for all the unknowns and uncertainties in the application, analysis, material properties, and manufacturing process. In some cases, the design margin is dictated to the engineer by policies or customer requirements. In other cases, the engineer must determine the appropriate design margin to apply based on the level of uncertainty during the design process and the level of risk that the customer is willing to accept. Some items that need to be considered when determining the design margin include confidence in the loading, ability of an operator to overload the equipment, level of analysis effort and confidence in the analysis technique, confidence in the material properties, fabrication and assembly processes/techniques, and environmental effects and potential for property damage, injury or death.

1.6 Environment

When designing any system or component, the engineer must understand the environment that it will be exposed to and how this environment may affect it. If the item is subjected to extreme thermal ranges, the design needs to accommodate the entire range. In some cases, the operating range is different from the storage or shipping range, so the design may have different requirements based on its state and/or configuration. The same can be said for other environmental conditions that the design may be exposed to such as rain, snow, ice, sand, and mud. In addition to direct loads placed on the design by its environment, any degradation of the design due to corrosion or other environmental conditions needs to be considered. Documents that can provide guidance for designing to environmental conditions include MIL-STD-810 and RTCA/DO-160. [Hempe 2011, 1–2]

As an actuator's environment can greatly influence the design and verification/validation process, the driving environmental conditions need to be clear at the start of a project. Some conditions that need to be considered in material and coating selection include maintaining acceptable properties through the temperature range, ability to withstand wear conditions, electrical and thermal conduction and expansion characteristics, and ability to withstand the environmental rain/snow/ice/saltwater/wash down requirements. The materials that satisfy these requirements then need to be evaluated and configured to satisfy other performance, weight, and cost goals. Some conditions that influence force/torque/power requirements include wind, snow, ice, base platform motion (as influenced by mobility loads, sea state, etc.), and altitude. As the reader can see, generating a complete set of environmental requirements with specific values can help the high-performance actuator engineer create a system that meets customer needs with the fewest design iterations.

1.7 Component Strength

Designing a component with adequate strength is one of the most important functions that an engineer can perform. The world depends on engineers to keep them safe from component failure. The process for designing adequate component strength includes determining the loads, constraints, initial overall product configuration, and then selecting materials and performing an initial structural analysis. Based on this initial analysis, component sizes and configurations can be modified to achieve the safety margin desired. As the design matures, the engineer needs to continue refining the analysis and all supporting data to make sure that they truly represent the final configuration, loads, material(s), and environment.

1.8 Component Stiffness

One often overlooked, but important aspect of actuator design is designing for overall product stiffness. This is especially true for high-performance systems where product weight and actuator speed is important. If stiffness is low, then the system natural frequency can be low and be excited during system operation. The natural frequency is given by the equation

$$f_n = \sqrt{\frac{k}{m}} \tag{1.1}$$

where f_n = natural frequency (also known as ω_n), k = stiffness, and m = mass. [Shortley and Williams 1971, 245]

If the system is excited during operation, it can become uncontrollable or unstable causing damage to itself, payload, or personnel. In many cases, stiffness can be the driving factor in material selection or design configuration.

1.9 Reliability

Owners and users expect their equipment to perform within its operational and life parameters. Successful products meet and often exceed customer expectations. One key measure of how well a product is going to meet expectations is through a reliability assessment during product design. The greatest impact on product reliability can be made during product design. By predicting product reliability with a structured reliability program, design improvements that target low reliability features or components can be implemented. Some key measures of reliability include mean time between failures or mean cycles between failures. There can be variations on these such as repairs per hundred/thousand/... hours/units/.... More details on reliability calculations are presented in Section 3.5.5. As the design proceeds through its development cycle, the validation testing should include steps that demonstrate that the product is on a path to meet its stated reliability goals. As with all other requirements, as the product reaches its final production, launch date data should show that they are on a glide path to meet its reliability requirements.

1.10 Maintainability

It is important to consider product maintenance early in the design process because that is when the most effective maintenance approaches can be implemented. Any periodic maintenance that can be designed out of a system reduces operating cost and downtime. In addition, minimizing periodic maintenance also minimizes the possibility of maintenance induced failures. Improving ease of maintenance and access for maintenance actions improves the probability of maintenance being done as recommended and maintainer satisfaction and again reduces maintenance induced failures.

1.11 Cost

Product cost has three primary components: initial procurement cost, maintenance and operating costs, and decommissioning costs. All of these are combined in the term life cycle cost. It is important to know the product's customer to understand if any single cost category is more important or if the customer is focused on the total life cycle cost. Sometimes it is easy to focus on initial product cost because the engineer just needs to focus on component, assembly and test, and marketing and distribution costs. Determining maintenance cost is more difficult because maintenance items need to be identified along with their procurement and labor costs. Operating costs require the engineer and cost analyst to identify product consumables and determine consumable costs along with any operator labor costs. Automated machinery may have lower operating costs because labor costs are minimized (this is a significant reason for development of new equipment with high performance actuators). Finally, decommissioning costs occur at the end of the products useful life. This area is becoming more important as the cost of disposing hazardous materials becomes more expensive and recycling becomes more prevalent. If the engineer can avoid using hazardous materials in the product, cost associated with hazardous materials can be avoided in all cost categories and utilizing recyclable materials increases the salvage value at the end of the product's life. As discussed here, product design can drastically affect cost and it is important to understand the customer's cost sensitivity.

1.12 Summary

Like any other part of mechanical design, successful actuator design needs to consider performance requirements, applied loads, load reactions or constraints, and design safety margin required, along with operating and storage environments to design an actuator with appropriate strength and stiffness. Other factors to consider include reliability, maintainability, initial cost, maintenance costs, and decommissioning or scrap costs. One can design an actuator that operates with amazing speed in a wonderful package, but if it is unreliable or cannot be maintained, it will ultimately be rejected by customers. Likewise, if any of the cost factors is too high, the product will not be a success. The general development process used is defined in Figure 1.1. ANSI/EIA-632, ML-STD-499, Naval Systems Engineering Guide, or systems engineering processes provide a proven standardized framework for development programs or projects. [Naval Air Systems Command, et al. 2004, vi], [Leonard 1999, 3]

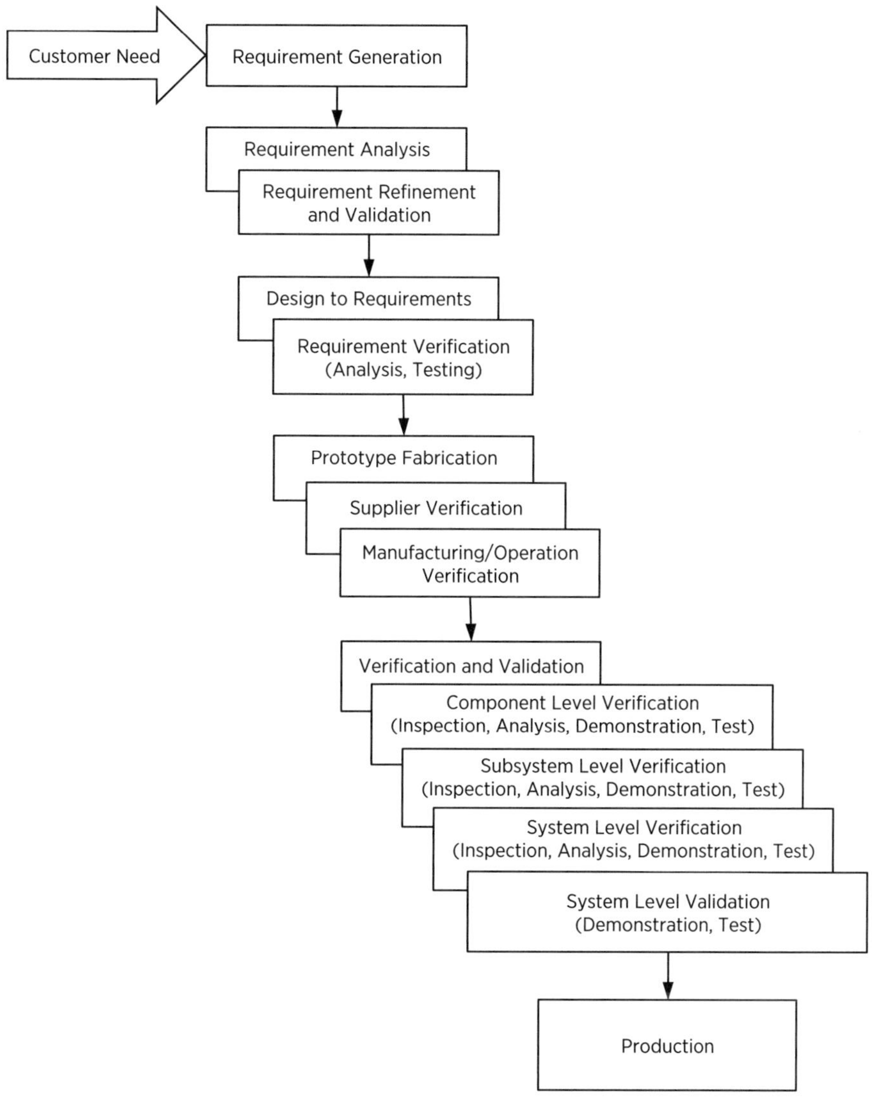

Figure 1.1 Development process.

1.13 References

Shortley, G., and Williams, D. 1971. Elements of Physics, Fifth Edition. Prentice-Hall, Inc: Englewood Cliffs, NJ.

Naval Air Systems Command, Naval Sea Systems Command, Naval Supply Systems Command, Space and Naval Warfare Systems Command, Marine Corps Systems Commnad. 2004. Naval Systems Engineering Guide. Department of the Navy: Washington, DC.

Hempe, D. W. 2011. Advisory Circular 21-16G. U.S. Department of Transportation, Federal Aviation Administration: Washington, DC.

John Leonard. 1999. Systems Engineering Fundamentals. Defense Systems Management College Press: Fort Belvoir, VA.

Chapter 2
Project Management

Project management for high-performance actuator systems is similar to that required for all other technology areas. It needs to establish and maintain the business case for the project. To do this, project scope needs to be defined and customer requirements need to be captured and maintained so that they can be decomposed into technical requirements that can guide the development process. An overall schedule that defined customer and market needs must be established so that detailed development schedules can be created. An entire picture of the costs associated with the project needs to be developed and maintained so that the development team and senior management understand the budget constraints that support the business case. An assessment of development risk also needs to be created, documented, and maintained so that progress against development challenges can be assessed by both the development team and senior management. Finally, the steps and items needed for final project close out needs to be defined so the project can be closed in a timely manner.

2.1 Scope

One of the first items that need to be accomplished by the project manager or project management team is to define the scope of the project. Everything that is in scope needs to be defined. This includes market(s) to be served, applications and/or platforms to be supported, configurations to be developed, and future growth areas. It is equally important to define the items that are out of scope for the project. Items in this list may echo the "in scope" item topics, but clearly define which features are not going to be considered in the project. This clearly defines the scope of work so there is no confusion between the development team and management.

2.2 Requirements

Requirements that define market and customer needs must be generated to guide the development process and define when the development goals have been reached. Good requirements should be specific, define the need, and have clear acceptance criteria. At the project management level, the key performance parameters or key requirements that shall be met need to be defines. These requirements can then be decomposed to define requirements specific to the development team(s). The project may follow a defined systems engineering approach where the market or customer need is transformed into a final set of requirements that define the operational parameters and configuration. These requirements should incorporate all the technical, reliability, maintainability, safety, and human factors into the development effort so that the final product represents an integrated solution, not a collection of individual solutions. Development and maintenance of the requirement set is collaboration between all members of the development team(s) and management. Some of the popular methods to help develop requirements include KJ diagrams or affinity diagrams, content analysis, acquisition assessment, capability assessment, and interviewing. [Naval Air Systems Command et al. 2004, 68–83], [Monarch, Goldenson and Stoddard 2009, 4–7], [Ulrich 2003, 2]

2.3 Schedule

A comprehensive schedule capturing all the milestones for the project needs to be generated as part of the project management area. This schedule should be generated in concert with the extended team including marketing, design engineering, manufacturing, purchasing, reliability, service, and finance. Each of these functions needs to define the parameters of parameters important to them so that the interdependencies can be captured. Using this information, the relationships between each function can be defined and integrated into the plan so that deliveries between the various functions can be established along with the required timing. The collection of the milestones, the time to achieve them, and interdependencies should form the project master schedule. The master schedule should define the time to market for the product and may include a phased approach that defines when each product configuration becomes available to support the market need. The master schedule supports an assessment of how well the plan supports market needs and anticipated competitor product releases. If the master plan shows that the product may be released too late to satisfy the business plan, it needs to be reviewed to see if opportunities exist to accelerate development. The risks associated with an accelerated development plan then need to be evaluated. The master schedule should also integrate development costs by phase so that a project spend plan can be developed. If the plan does not align with available resources, financial (development funding), design (engineers, analysts, etc.), test (test engineers, test facilities, test equipment, etc.), manufacturing (manufacturing engineers, prototype manufacture, raw material availability, manufacturing space, manufacturing equipment, etc.), purchasing (buyers, supplier quality personnel/representatives, approved supplier base, etc.), and other critical project personnel, then the plan needs to be modified to satisfy the resource constraints. Again, this modified plan needs to be reviewed to make sure that it continues to support the business case. [Naval Air Systems Command et al. 2004, 32–35], [MIL-STD-499 1969, 8–10] A sample project development schedule is shown in Figure 2.1.

No.	WBS	Task	Duration (Days)	Start	End	Resource(s)	Predecessor(s)
1	1.0	Actuator System	0	1/1	1/1	PM	
2	1.1	Hardware	0	1/1	1/1	PM, Eng	
3	1.1	Project Definition	15	1/1	1/20	PM, Eng	
4	1.1	Requirement Definition	15	1/20	2/8	PM, MKT, Eng	3
5	1.1	Requirement Analysis	120	2/8	7/25	Eng	4
6	1.1	Concept Generation	80	3/21	7/3	Eng, Anal	5
7	1.1	Concept Analysis	50	4/4	6/11	Eng, Anal	6
8	1.1	Requirement Review	1	7/25	7/28	PM, Eng	5
9	1.2	Preliminary Design	120	7/29	1/13	Eng	8
10	1.2	Preliminary Design Analysis	60	9/9	11/30	Eng, Anal	9
11	1.2	Preliminary Design Testing	50	9/23	11/30	Eng, Test	10
12	1.2	Preliminary Design Review	1	1/13	1/16	PM, Eng	9,10,11
13	1.3	Preliminary Design	120	7/29	1/13	Eng	8
14	1.3	Preliminary Design Analysis	60	9/9	11/30	Eng, Anal	13
15	1.3	Preliminary Design Testing	50	9/23	11/30	Eng, Test	14
16	1.3	Preliminary Design Review	1	1/13	1/16	PM, Eng	13,14,15
17	1.2	Detail Design	120	1/17	7/4	Eng	12
18	1.2	Detail Design Analysis	60	2/28	5/21	Eng, Anal	17
19	1.2	Pre-Prototype Fabrication	50	3/14	5/21	Purch, Mfg	18
20	1.2	Detail Design Testing	40	5/21	7/14	Test	19
21	1.2	Detail Design Review	1	7/4	7/7	PM, Eng	17,18,19,20
22	1.3	Detail Design	120	1/17	7/4	Eng	16
23	1.3	Detail Design Analysis	60	2/28	5/21	Eng, Anal	22
24	1.3	Pre-Prototype Fabrication	50	3/14	5/21	Purch, Mfg	23
25	1.3	Detail Design Testing	40	5/21	7/14	Test	24
26	1.3	Detail Design Review	1	7/4	7/7	PM, Eng	22,23,24,25
27	1.2	Production Readiness	120	4/11	9/26	Mfg	17,19
28	1.2	Procurement	60	7/18	10/8	Purch	21
29	1.2	Fabrication	80	7/18	11/7	Mfg	21,28
30	1.2	Assembly	30	11/7	12/17	Mfg	28,29
31	1.2	Testing	40	12/17	2/9	Eng, Test	30
32	1.2	Product Verification	30	2/9	3/21	Test	31
33	1.3	Production Readiness	120	4/11	9/26	Mfg	22,24
34	1.3	Procurement	60	7/18	10/8	Purch	26
35	1.3	Fabrication	80	7/18	11/7	Mfg	26,34
36	1.3	Assembly	30	11/7	12/17	Mfg	34,35
37	1.3	Testing	40	12/17	2/9	Eng, Test	36
38	1.3	Product Verification	30	2/9	3/21	Test	37
39	1.1	Hardware Integration, Assembly & Test	20	3/21	4/16	Eng, Test	32,38
40	1.1	Product Verification	30	4/16	5/26	Test	39
41	1.0	Actuator Integration, Assembly & Test	20	5/26	6/21	Eng, Test	40
42	1.0	Product Verification	30	6/21	7/31	Test	41
43	1.0	Product Validation	40	7/31	9/23	Test	42

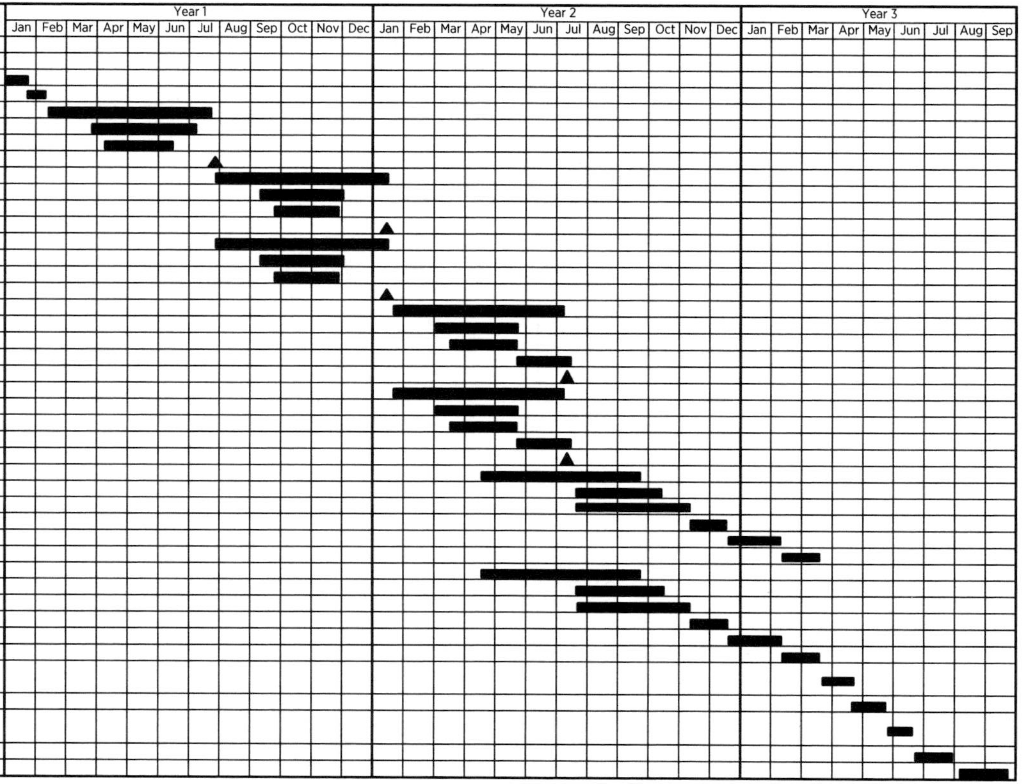

Figure 2.1 Sample project development schedule.

A work breakdown structure (WBS) may be used to define the products to be developed and delivered for the project. A WBS should relate all deliverables for the project such as hardware, software, support, and service to each other and the end product. WBS elements are listed and defined in a WBS dictionary that is prepared as part of WBS development. For each WBS element, the dictionary should show its relationship to other elements, the resources, and processes to produce it, the basic technical characteristics of the element, and linkage to the detailed technical description for the WBS element. The WBS can then be used to plan, budget, and schedule project work. [MIL-STD-881 2011, 8–16] With a plan, budget, and schedule in place, an earned value management system (EVMS) can be established to define project baseline cost and schedule objectives and track these objectives against actual cost and schedule values. Tracking cost and schedule variance provide an objective measurement of project performance. A simple measures to characterize project schedule performance with schedule performance index (SPI). With this values leadership can get a quick idea of how well a project is doing. An SPI of 1.0 indicates that the project is exactly on schedule. If the SPI is less than 1, the project is behind schedule and an SPI greater than 1 indicated the project is ahead of schedule. There are many other intricacies to the development of a WBS and implementing a full EVMS than covered here. Full books and courses in the WBS and EVMS development are available. This book is mainly concerned with introducing basic concepts to support development of an actuator (Figure 2.2).

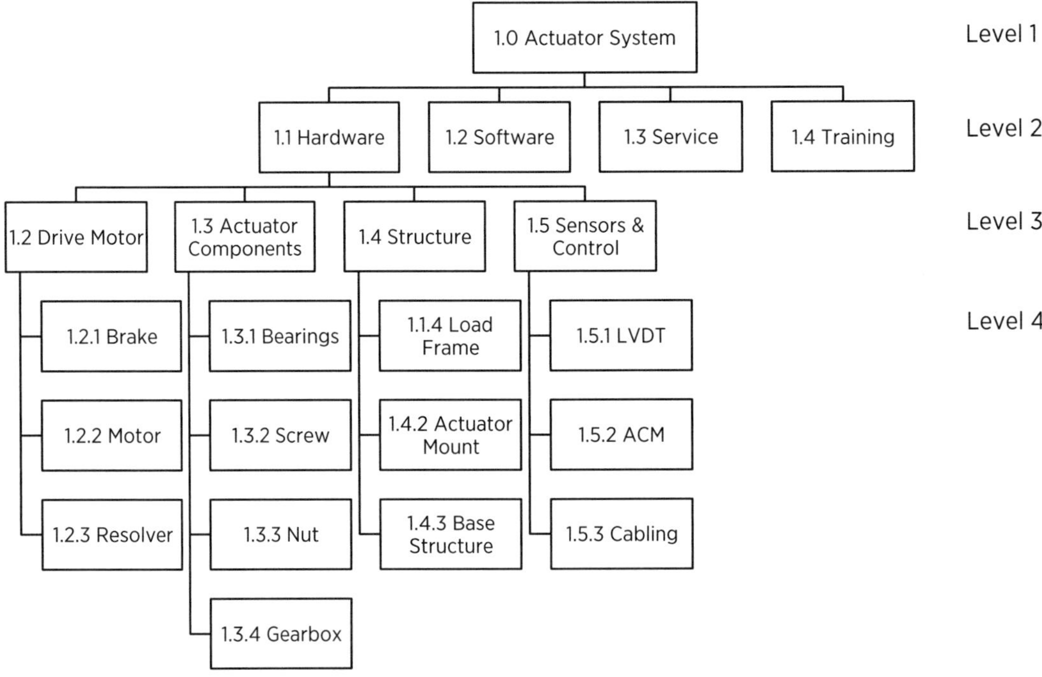

Figure 2.2 Sample actuator system WBS.

$$SPI = \frac{BCWP}{BCWS} \qquad (2.1)$$

where BCWP = budgeted cost of work performed and BCWS = budgeted cost of work scheduled. [Defense Contract Management Agency 2012, 18–19]

2.4 Cost

From the beginning of the design process until production ends, people always want the price to be lower. Most development projects go through five phases, design, prototype, production, operating, and decommissioning. Within each of these phases, others may exist. For example, the design phase may have requirements analysis, concept development, preliminary design, and detailed design phases. The prototype phase may have fabrication, assembly, test, checkout, and design refinement phases. Production may include final design, cost reduction, production tooling, design refinement, and production support phases. Operating costs include all costs associated with maintenance, energy usage, downtime, and operator labor. Finally, decommissioning includes all costs associated with taking the machine out of service and safely disposing of all its components. In all cases, people look to the engineer for input into how much each phase is going to cost and how long in terms of calendar time it is going to take. Early in the process, the team concentrates on cost activities associated with requirements analysis and concept development. Generally, the customer is looking for an estimate of all costs from this point through decommissioning. The greatest interest is usually in design, prototype, and production costs. The customer wants to know if the design is affordable and is eager to get a prototype built for two main reasons. The first is that once a working prototype exists, evidence that the key requirements can be met is available and the second is that an actual machine is available show external customers. This also helps gage customer acceptance and lets the sales organization start judging how much customers are willing to pay. Along with this, internal customers need to know how much the machine is going to cost so that they can judge market acceptance, price thresholds, and potential profit margins.

Important considerations that need to be managed to keep project cost under control are scope changes, project delays, and unanticipated technological challenges. During development, it can seem easy to implement new features or specific design features that a customer wants, but each of these need to be evaluated in terms of development and the final product cost. While it is important to satisfy customers, the design and management teams need to evaluate the potential for these items to drive development and product cost up. If the additional cost can be justified in a solid business case, then the features should be incorporated. If not, then the best strategy may be to complete the design as planned and offer the additional features as options or upgrades at a later date. Sometimes project due dates are delayed by factors outside the design community. Again the design and management teams need to evaluate if they want to delay the program or keep it on the existing plan. If the resources cannot be reduced for a delayed product release, then the overall project cost is going to increase, potentially without gaining any new features or product cost reduction. This just reduces the overall value of the project. Finally, additional development work may need to be completed as part of the project because a technology that it depended on did not mature as expected. This type of extra work needs to be carefully evaluated to make sure the team, subcontractor, and supplier have a high probability of success at the new anticipated cost. If not, the team should seek alternatives and develop a contingency plan to maintain project feasibility. How well the project is performing relative to cost can be easily captured by calculating its cost performance index (CPI). A CPI of 1.0 means the work accomplished is exactly equal to the work planned, while a CPI less than 1 indicates that the project is costing more than planned and a CPI of greater than 1 indicates that the project is costing less than planned.

$$\mathrm{CPI} = \frac{\mathrm{BCWP}}{\mathrm{ACWP}} \tag{2.2}$$

where BCWP = budgeted cost of work performed and ACWP = actual cost of work performed. [Defense Contract Management Agency 2012, 15–16]

2.4.1 Design Cost

Actuator designers are not the only ones that need to help estimate, track, and control design costs. However, complexity and risk increases as performance goals increase because the state of the art is usually being challenged. If the design is similar to a previous effort, then design costs can be accurately estimated from historical costs. For other design efforts, the best approach is to factor costs from a previous design. If detail historical cost information is available, calculation of design cost for a new program may take the following form:

$$\text{Design Cost} = \mathrm{DS} \times C \times P \times \mathrm{DA} \times \mathrm{TD} \times S \times R \times \mathrm{HDC} + \mathrm{LD} \times \mathrm{HDocC} + \mathrm{MR} \times \mathrm{HMRC} \tag{2.3}$$

where DS = design scope factor

C = complexity factor

P = performance factor

DA = design approach factor

TD = technology development factor

S = schedule factor

R = risk factor

HDC = historical design cost

LD = level of documentation factor

HDocC = historical documentation cost

MR = management and review factor

HMRC = historical management and review cost

If detailed historical cost information is not available, calculation of design cost for a new program may take the following form.

$$\text{Design Cost} = \text{DF} \times \text{DC} + \text{AF} \times \text{AC} + \text{DocF} \times \text{DocC} + \text{MRF} \times \text{MRC} + \text{MR} \quad (2.4)$$

where DF = design factor

DC = design cost estimate

AF = analysis factor

AC = analysis cost estimate

DocF = documentation factor

DocC = documentation cost estimate

MRF = management and review factor

MRC = management and review cost estimate

MR = management reserve

Performing a "bottoms up" cost estimate without the benefit of previous project costs usually ends up with an optimistic figure. A "bottoms up" estimate is where the estimator prepares a list of design activities that they think are necessary and then assigns a level of effort to each activity. Unfortunately, the list is usually incomplete and/or the scope of each activity is underestimated because the current state of the design is overestimated or the estimator does not have the experience to understand the level of effort required to mature the design. It is important to track design progress and costs, then make an estimate on the level of effort to complete the design. The estimate should include both cost and schedule. This information gives the engineer an idea of how the design is progressing, if it can be completed within the allocated budget and if it can be completed as scheduled. Upper management and design customers are also interested in this information, so it should be passed on so that they understand how the design is progressing. Good cost and schedule information is always welcome, but communicating cost and/or schedule overrun information is just as important. A plan to deal with overruns needs to be put in place before they get out of control and drastic steps such as canceling the design need to be taken.

2.4.2 Prototype Cost

Estimating the cost of fabricating, assembling, and testing a prototype is much the same as estimating design costs. Data from previous programs is the best source of information. Building and testing high-performance actuators is different from structural components due to the level of precision required to make them work accurately and smoothly. Building a prototype can also be different from building production hardware because the fabrication and assembly techniques may be different. For example, the engineer may choose to machine a component out of a piece of wrought material for the prototype instead of making the casting planned for production units. This allows the design to be proved without pattern schedule and cost impacts. However, the engineer needs to make sure that such substitutions represent the anticipated production configuration and any differences that exist are not significant. For example, the difference between wrought and cast material properties needs to be evaluated for its impact on test goals. The program may also want to save the expense of complete assembly tooling, but some tooling may be required to assemble the prototype. This tooling may include items such as aids to assemble special seals, bearings, and springs. Such aids are good investments because the components may be expensive, have a long lead time, and/or be difficult to assemble without tooling. Actuator testing generally requires test hardware not needed for simpler mechanisms. Test hardware may include instrumentation to assess performance levels, separate power supplies, and separate control systems to control actuator motion. The test setup should ensure safety by including previsions for managing a runaway actuator so that it does not harm personnel, surrounding equipment, or itself. All these items are usually not fully defined at the beginning of the program when budgeting for the program is done, so the design engineer is in the best position to provide this data the customer. Prototype cost and schedule data should be updated as the design matures, so everyone is confident adequate funds and time are available to complete the actuator effort. If the design is similar to a previous effort, then prototype costs can be accurately estimated from historical costs. For new hardware, the best approach is to factor costs from a previous design. If detailed historical cost information is available, calculation of prototype cost for a new program may take the following form:

$$\text{Prototype Cost} = F \times C \times P \times \text{DA} \times \text{TD} \times S \times R \times \text{HPC} + \text{LoT} \times \text{LD} + \text{HTC} + \text{MR} \times \text{HMRC} \quad (2.5)$$

where F = fabrication factor

C = complexity factor

P = performance factor

DA = design approach factor

TD = technology development factor

S = schedule factor

R = risk factor

HPC = historical prototype cost

LoT = level of testing factor

LD = level of documentation factor

HTC = historical test cost

MR = management and review factor

HMRC = historical management and review cost

If detailed historical cost information is not available, calculation of prototype cost for a new program may take the following form:

$$\text{Prototype Cost} = \text{FF} \times \text{FC} + \text{AF} \times \text{AC} + \text{TF} \times \text{TC} + \text{DF} \times \text{DC} + \text{MRF} \times \text{MRC} + \text{MR} \quad (2.6)$$

where FF = design factor

FC = fabrication cost estimate

AF = assembly factor

AC = assembly cost estimate

TF = test factor

TC = test cost estimate

TSF = test setup factor

TSC = test setup cost estimate

DF = documentation factor

DC = documentation cost estimate

MRF = management and review factor

MRC = management and review cost estimate

MR = management reserve

As with design cost, performing a "bottoms up" prototype estimate without the benefit of previous project cost and schedule usually ends up with optimistic figures. It is important to track prototype progress and costs and then make cost and schedule estimates for the level of effort to complete the prototype. This information gives the engineer, customer, and management an idea of how the prototype is progressing, if it can be completed within the allocated budget and if it can be completed as scheduled.

2.4.3 Production Cost

Estimating the production cost of an item is much like estimating the prototype cost. At the beginning of a design project, someone has determined there is a market for the product and it can be sold to the consumer for an acceptable price. The design engineer is going to be asked to verify this. It is important for the engineer to understand the consumer's ability and willingness to pay for the features of the product. As the engineer goes through the production cost estimating process, notes should be made on cost drivers and items that can be removed to reduce cost. Likewise, items and features that can be added to the product with little or no cost impact should be captured. The main difference between prototype cost and production cost lies in the quantity produced. For very low production quantities or production rates, there may not be a large difference in cost. If detailed historical cost information is available for similar products, calculation of production cost for a new product may take the following form:

$$\text{Production Cost} = C \times P \times \text{PR} \times \text{PQ} \times T \times \text{Test} \times \text{WF} \times \text{LPS} \times \text{OH} \times \text{HPC} + \text{OC} \tag{2.7}$$

where C = complexity factor

P = performance factor

PR = production rate factor

PQ = production quantity factor

T = tooling factor

Test = test factor

WF = warranty factor

LPS = level of production support factor

OH = overhead factor

HPC = historical production cost

OC = option cost

If detailed historical cost information is not available, calculation of the production cost for a new product may take the following form:

$$\text{Production Cost} = (PF \times PC + MF \times MC + AF \times AC + TF \times TC) \times WF \times SF \times OH + OC \quad (2.8)$$

where PF = purchased part factor

PC = purchased part cost estimate

MF = manufactured part factor

MC = manufactured part cost estimate

AF = assembly factor

AC = assembly cost estimate

TF = test factor

TC = test cost estimate

WF = warranty factor

SF = support factor

OH = overhead factor

OC = option cost

The engineer along with the customer has some flexibility in determining production cost by adding or removing features from the final product. Production cost can also be influenced by production methods and level investment in tooling or fixtures. The level of investment in reducing production cost needs to be evaluated with the number of units to be produced so that the optimum level of investment can be determined. It is important to accurately determine the anticipated minimum production cost and compare it with the maximum price the customer is willing to pay. If the two start approaching each other, the feasibility of the product needs to be examined. When this happens there are three choices: reduce the production cost, raise the price, or cancel the project. The first two are often at odds with each other because the first usually removes features, while the second adds or maintains features. However, there are times when insignificant features have high production costs associated with them and features worth a lot to the customer do not. If these can be identified, the product feature list can be modified; otherwise, the third option may need to be implemented. Production cost goals may be included in the design requirement set.

2.4.4 Operating Cost

Estimating the operating cost of an item requires estimating how often maintenance is required, how long maintenance is going to take, the cost of maintenance materials and labor, energy and other items consumed during operation, and the cost of downtime. This information is often desirable at the beginning of a design project so that the customer and marketing representatives can make sure the design is going to be affordable. With the cost of energy and resources continuing to increase, engineers are being asked to minimize their use. Therefore, as the engineer goes through the operating cost estimating process, notes should be made on how operating costs can be reduced. Features that can be added to reduce operating costs should be considered and listed as possible options for the customer. This approach lets the customer decide the value of reducing operating costs for their specific operation. Calculation of operating costs may take the following form:

$$\text{Operating Cost} = \text{EC} \times E + \text{CC} \times C + \text{MLR} \times \text{ML} + \text{OLR} \times \text{OL} + \text{DT} \times D \tag{2.9}$$

where EC = energy consumption

E = energy cost

CC = consumable consumption

C = consumable cost

MLR = maintenance labor required

ML = maintenance labor cost

OLR = operator labor required

OL = operator labor cost

DT = downtime

D = downtime cost

The engineer along with the customer may be able to adjust the actual operating cost by adding or removing features from the final product. It is important to accurately determine the anticipated operating cost and compare it with current machine costs. In addition to offering new capabilities, new machines should demonstrate reduced operating costs as a way to help pay for themselves. If operating cost is deemed a significant sales or customer benefit, it should be included in the requirement set.

2.4.5 Decommissioning Cost

Years ago, people did not pay much attention to decommissioning costs. When a products life was over, it was scrapped and replaced by a new one. Now, after spending billions of dollars on cleaning up pollution from sites where old equipment was scrapped, officials are more sensitive to how a product is replaced at the end of its life. In the most favorable scenario, the product does not have any hazardous materials and has a lot of materials with a high recycling value. Dealing with hazardous materials in a product at the end of its life adds to the overall cost of ownership and has become such a significant factor that using some materials in the product are now prohibited though laws or the initial requirement set. Prohibited materials need to be carefully reviewed during the requirements analysis phase because alternative materials can add to design complexity or production cost. On the other hand, using materials that are easy to recycle and that have high recycle value adds to the value of the machine at the end of its life, essentially reducing the cost of ownership. The decommissioning cost of a product may take the form:

$$\text{Decommissioning Cost} = \text{HMC} \times \text{HMM} - \text{RMC} \times \text{RMM} + \text{DAC} \tag{2.10}$$

where HMC = hazardous material cost

HMM = hazardous material mass

RMC = recycled material cost

RMM = recycled material mass

DAC = disassembly cost

2.4.6 Cost Estimating Techniques

There are many techniques and tools available to engineers for estimating program costs. Some are as simple as telling the customer or management how much you think the job is going to cost. This may be satisfactory if the project is small or going to be reviewed in depth at a later date for another gate review when more details are available. However, it is rarely good enough for an organization to commit significant financial, labor, and facility resources. Today, most organizations use some sort of cost estimating model to determine development program cost and then an EVMS to track progress against the plan. Depending on the model, the process may be as complex as defining individual components, entering them into a program that determines all costs associated with their development, production, integration into the final product, and decommissioning of the product. To support this effort the team may need to gather the cost of standard commercial components, which while tedious, is usually reliable, consistent and easy to obtain from a supplier. Other components reflecting custom manufacture and design intent are harder to estimate without insight into previous production history. Most companies protect their cost information because it is considered propriety and contributes to their advantage over competitors. The same is true for assembly and test processes. Complex programs are available to help estimate the design, prototype, assembly, test, and production costs, but they are usually reserved for use on very large development programs. Such programs, once calibrated, provide good cost data and their data therefore instill a lot of confidence from leadership. They can also be used to update a project's estimate to complete and the resulting estimate at complete. [Defense Contract Management Agency 2012, 17, 22] As with all such programs, the results need to be reviewed to make sure they make sense. If they do not, the input data should be reviewed to understand factors influencing the results. Once calibrated, this data can then be utilized as the design matures and updated as necessary:

$$\text{EAC} = \text{ACWP} + \text{ETC} \tag{2.11}$$

Costs can be separated into three distinct phases with each phase also having individual cost categories. The total cost breakdown for a product may be represented as follows:

1. Design
   a. Requirements analysis/concept formulation
   b. Preliminary design
   c. Detail design
   d. Prototype fabrication
   e. Prototype assembly and test
2. Production
   a. Production design
   b. Production support
   c. Production facilitization and tooling
   d. Raw material
   e. Purchased components
   f. Manufactured components
   g. Assembly
   h. Test
   i. Shipping
3. Decommissioning
   a. Disassembly
   b. Hazardous material management
   c. Recycle materials and components
   d. Restore facility

Design engineers are not responsible for all these costs but as they are creating the design, they know the most about the product, so their input is needed to get the best cost data. This is rarely the engineer's favorite task but nonetheless part of an engineer's job.

2.5 Risk

Development programs always involve risk. Successful programs understand the associated risks and actively manage them. Identification and management of risk requires risk assessment and tracking and risk reduction programs. Some types of risk include schedule, cost, technology maturation, and customer acceptance.

2.5.1 Schedule

Schedule risk involves the ability to get the product on the market as planned. Schedule delays usually increase cost because personnel are charging to product development for a longer period of time. It also means the product is getting to market later; delaying profits and giving competitors a chance to catch up or get further ahead. Schedule risk is monitored by assessing progress and making accurate estimates to complete the effort. The more resolution incorporated into the development plan, the more accurate these assessments and estimates can be. If the program has several parts to it, each part should be assessed separately to make sure that they are each progressing as planned. It is important to keep the parts synchronized so that integration of all parts at the end of the program can be made without delays or the need to substitute parts. Schedule adjustments can be made by adding or removing personnel or other resources as needed. However, it should be noted that all problems cannot be solved by just adding personnel.

2.5.2 Cost

Cost risk involves risk with all the items just covered in Section 2.4. Cost needs to be monitored, assessed, and estimated just as schedule does. However, cost problems cannot be fixed by adjusting personnel like schedule problems can. Unfortunately, cost growth is a real part of most development programs. Risk becomes unacceptable when development funds are no longer available or product profits are not able to cover its development. Cost management is covered in Section 2.4.

2.5.3 Technology

Most new products incorporate new technology; this is certainly true for high-performance actuators. Higher performance usually means pushing available technologies toward their limit or in extreme case developing new technologies. An assessment of the technology readiness level of all new technologies to be incorporated in the design should be completed. This step determines how mature the various technologies to be used are so a technology development plan can be implemented to control the risk associated with immature technologies. This plan needs to have adequate resolution into the individual technologies so that accurate assessments and estimates can be made against it. Just as schedule and cost are monitored, development and application of technologies utilized need to be assessed and estimates to complete technology maturation made. If these assessments and estimates do not show that technology development is on track, a risk mitigation plan needs to be implemented. This plan should have components that deal with improving the maturity rate of the primary technology plus plans for incorporating alternate technologies if the primary one does not work out.

2.5.4 Customer Acceptance

Finally, there is always risk that the customer may not accept the product. This can be for various reasons but is devastating for the company involved. The primary customer acceptance criteria should be contained in the design requirements set, but the engineer has some control over how the requirements are implemented and therefore product appeal to the customer. There are various tools and approaches to gathering voice of customer information. Some of them include interviewing, reviewing competitor information, and identifying trends (technological, demographic, financial, etc.). This information can be translated into a set of requirements to guide the development team through techniques such as a KJ Analysis, content analysis, or combination of them. [Monarch, Goldenson and Stoddard 2009, 4]

In some cases, an organization may be under contract to develop and deliver a product to a given set of requirements according to a specific development schedule. This approach can address customer acceptance and development schedule questions but may expose the organization to consequences if the final product cannot be delivered as promised. The agreement between the developing organization and customer, or contract, may contain penalty clauses for not meeting requirements or schedule. These need to show up in the key requirement set so that they have complete visibility to the development and management team.

The cost component of this risk area is monitored as cost risk and the schedule component as schedule risk. Other customer appeal factors should be carefully monitored during design and program reviews to make sure performance, aesthetics, applied technology, and business case are on track to deliver as promised. Any concerns that arise during these reviews be recorded as issues or action items with responsible person and closure date assigned. This provides a mechanism for tracking their progress and timely identification of potential problem areas.

2.5.5 Risk Management

Risk is potential harm to the project or system under development. It can be thought of as made up of three parts: the things that can go wrong in or during the project, the consequences of these things, and the probability of them happening. Risk should be assessed for the project, end product, and processes associated with the system. Consequences of process variability should be included. Risk can come from technical sources (technology maturity, producibility, testability, usability and effectiveness of the end product), cost (development cost and end product cost, i.e., ability to develop within a positive business case), schedule (ability to hold the development schedule, development of supporting components, development of competitor products, customer need dates), and program (overall resource availability, alignment with organization goals). Risk management needs to identify problem areas, define an action plan to correct the problem, and identify the problems in a timely manner so that they can be addressed before severe damage is done.

Risk management should address the three risk elements identified earlier: items that can go wrong, consequences, and probability. The first can be addressed through design, manufacturing, service, procurement, customer and program reviews. Data gathered during these reviews should highlight the development status, resource availability, competitor status, and customer need as related to each of these areas. This list of items can then be assessed for areas of concern and the consequences of the concern. Each of these should be scored so that a ranking of concerns can be assembled. Along with this, the probability of each concern happening should be identified. With this information, a risk mitigation plan can be assembled based on the effect of each concern and the probability of it happening. The risk mitigation plan needs to identify methods to regularly assess each concern and include actions to address them. Some of these actions may be implemented as part of the development program if the risk has a high rating or may be just available for implementation when review of a concern indicates that progress has gone offtrack. The risk mitigation plan should be reviewed regularly to close out concerns that have been fully addressed, add new concerns identified, and identify existing concerns that have gone off-track. Risk management should increase program stakeholder confidence and lead to successful program completion. [Naval Air Systems Command et al. 2004, 158]

2.6 Project Close Out

As the project moves through the various development stages, the project manager or project management team (depending on project size) should capture the lessons learned, ensure project documentation is kept up to date, and update senior management on new opportunities that come from the project.

Lessons learned include everything learned from customers, the design team, analysis and test, marketing, and management. Requirements can be captured in the requirement set, but the team may also learn about new customers, how customers like to be treated, and valuable sources of information within the customer's or end user's organization. New analysis techniques may be developed or validation of analysis models may occur that as part of the project. The marketing team may develop new sources of information or an understanding of a new market. Project management may develop new techniques and approaches that improve overall project efficiency. All this information should be recorded in a standing record that is published as part of project close out.

As the design and project mature, new documentation is created to represent the latest configuration, plan, and business case. Along with the new documentation, an obsolescence plan should also be created to manage all the documentation being replaced. This plan records the intended implementation of the design improvements, plan, and supporting business case so that an orderly transition from the old state to the new one can be communicated throughout the organization. This plan needs to define the obsolescence for the previous version(s). Without a clear plan the team becomes burdened with maintenance of all the data and the project runs the risk of using on an old version when a more recent one is desired.

Finally, project close out should address how opportunities created during the project are going to be addressed. It should define if the opportunities become an extension of the existing project or become a new project with its own business case. Without such a plan, new features or products may be added to the existing project so that it never ends or they may be lost and never implemented.

2.7 References

Naval Air Systems Command, Naval Sea Systems Command, Naval Supply Systems Command, Space and Naval Warfare Systems Command, and Marine Corps Systems Commnad. 2004. Naval Systems Engineering Guide. Department of the Navy: Washington, DC.

Ulrich, K. 2003. KJ Diagrams. The Wharton School, University of Pennsylvania: Philidelphia.

Commander, Air Force Sytems Command. 1969. "MIL-STD-499, System Engineering Management." www.everyspec.com. (Accessed October 2015)

Monarch, I. A., Goldenson, D. R., and Stoddard, R. W. 2009. An Innovative Requirements Solution: Combining Six Sigma KJ Language Dage Analysis with Automated Content Analysis. SEPG 2009 North America: San Jose.

U.S. Department of Defense. 2011. MIL-STD-881, Work Breakdown Structures for Defense Material Items. U.S. Department of Defense: Washington, DC.

Defense Contract Management Agency, Department of Defense. 2012. Earned Value Management System (EVME) Program Analysis Pamphlet (PAP). Defense Contract Management Agency, Department of Defense: Virginia.

Chapter 3
Requirements Analysis

One of the first steps in designing a high-performance actuator is to determine the requirements it has to meet, including all items covered in chapters 1 and 2. For a high-performance actuator, significant items that differentiate it from a standard actuator and needs to be assessed include payload, move time or velocity, and range of motion requirements. After that, the designer can move on to assess structural loading, design margin, environmental effects and constraints, reliability, maintainability, and cost requirements. During this phase of the design process, each requirement influencing the product must be carefully evaluated to make sure a feasible solution exists to address all requirements. This phase should also include a requirement validation step to insure there is no redundancy or conflicts between requirements and they connect to ensure that the end product delivers the customer's needs and expectations. This means, at a minimum, that an initial design concept with a validated set of written requirements is available for review at the end of the requirements analysis. This sets the foundation for system verification where an assessment of the design is completed through inspection, analysis, demonstration, and/or test to ensure that it satisfies the requirement set. Successful completion of these steps should ensure passing end product validation activities where the product demonstrates that it satisfies the customer's validated requirement set.

Requirements analysis includes design, design analysis, and preprototype testing steps necessary to make sure that the design process is on track to deliver a product that can satisfy the top level requirements. These activities should ensure that the requirements accurately reflect customer key performance and environmental needs and expectations. As part of the requirement, validation process analysis and modeling should be conducted to ensure that the requirements can be met and no conflicts arise between requirements. Through the design, analysis, and test activities, this phase also provides the opportunity to flow top-level requirements to the design teams and decompose the requirements. These decomposed and derived requirements can then effectively guide design of lower level systems, subsystems, and components. Some key contributors to the design of a high-performance actuator that support a structured development program are covered in the following sections. Section 3.1 covers actuator motion and the related basic physics principles. In the section on loads, the factors that need to be considered as loads to support the concepts presented in Section 3.1 are introduced. As component stiffness can be a significant factor in determining the performance level of an actuator, it is covered in a separate section. The section on strength outlines the design considerations needed to ensure that product function, life, and reliability goals are reached. As how an actuator and its various components are constrained influence both stiffness and strength, they are covered in a separate section. Finally, verification and validation, as they relate to this early stage of the development process, are discussed. [Naval Air Systems Command et al. 2004, 1]

3.1 Performance

For actuator design, performance requirements come from the need to move a payload through some range of motion or achieve a certain velocity within a given time. At this point, it is useful to apply the concepts of conservation of work, impulse, momentum, energy and power to analysis of the actuator(s). Utilizing these principles, the engineer can make sure nothing is missed and ensure that the analysis and modeling of system behavior matches real-world performance. Of course, the design needs to compensate for friction losses and other inaccuracies before the analysis can match real world performance. However, ignoring these losses during initial analysis allows the analysis to be easily checked for conservation principles, so this is often a good place to start.

As the design matures and the design engineers refine the analyses, they need to make sure that the boundary conditions that drive each is considered. For example, when utilizing a hydraulic system as the power source, the actuator operation at minimum and maximum operating pressures should be calculated or simulated. One can clearly see how system pressure fluctuates if hydraulic fluid usage similar to that shown in Figure 5.1 is assessed with predictions of accumulator volume, pump response, flow rate, temperature, and so on effects incorporated. Such an assessment or system model can help predict actuator performance over the range of operating conditions that should include environmental, orientation, and loads. One might even need to assess operating in a degraded mode. Similar considerations need to be made for pneumatic and electric drives where operating temperature, pressure, and voltage ranges can influence system performance.

As we are considering high-performance actuators, designs should also be assessed for operational capability to design requirements due to friction, environmental effects, and payload changes. Friction effects can show up in numerous forms that all need to be considered. If not, actual performance may not match predicted performance and likely not meet the design targets. Some friction effects that should be considered are shown in Table 3.1. Environmental considerations that should be incorporated into the design assessment process are shown in Table 3.2. Finally, the payload may change or how it appears to the actuator may change over time or through the motion profile; some load effects that should be considered in the actuator design process are shown in Table 3.3.

Table 3.1 Friction considerations

Friction Source	Considerations
Seals	Static coefficient of friction
	Dynamic coefficient of friction
	Seal load when pressurized
	Seal load when unpressurized
	Seal friction when new
	Seal friction when broken
	Seal friction when worn
Rolling element bearings	Static friction
	Dynamic friction
	Friction as a function of preload
	Integrated bearing seal friction
	Lubrication type
	Axial load effects
	Radial load effects
Plain bearings	Radial load pressure (P)
	Shaft speed (surface velocity V)
	$P \times V$ value
	Seal friction
	Lubrication type
Power Screw	Seal fiction (static, dynamic)
	Type (roller, ball, acme, . . .)
	Nut preload
	Nut stiffness
Gear systems	Gear type (spur, bevel, spiral bevel, worm, planetary, . . .)
	Gear quality
	Backlash
	Ratio
	Preload
	Bearings
	Lubrication type
	Efficiency/thermal effects

Table 3.2 Environmental considerations

Environmental Effect	Considerations
Temperature	Part fit due to thermal expansion/retraction
	Change in part size due to thermal expansion/retraction
	Coefficient of friction changes with temperature
	Change in material modulus
	Change in material strength
Humidity	Water loss/gain (part shrinkage/growth)
	Change in material properties
Ice	Additional load
	Force required to break
	Changes in range of motion
	Damage due to expansion when freezing
Rain	Wet versus dry coefficient of friction values (static, dynamic)
	Additional load
Dirt/Dust	Clean versus dirty coefficient of friction values (static, dynamic)
	Additional load
	Increased wear

Table 3.3 Load considerations

Load Effect	Considerations
Payload change	Gain/loss
	Change in mass reflected to the actuator
	Change in inertia reflected to the actuator
	Change in compliance
Payload motion path	Change in effective ratio
	Change in mass reflected to the actuator
	Change in inertia reflected to the actuator
	Change in friction reflected to the actuator

If the effect of these changes and disturbances are successfully incorporated into the design through analysis and/or modeling, the initial prototypes have a good chance of operating as desired. Through a coordinated prototype test and an evaluation program, the designs and accompanying analyses can be improved. The final product should then reflect an optimized solution that meets all key design requirements.

3.1.1 Work

Work is the force applied over a distance. It has units of force × distance for linear systems or force × distance^2 for rotary systems

$$W = F \times x \text{ for linear systems} \tag{3.1}$$

$$W = T \times \theta \text{ for rotary systems} \tag{3.2}$$

where W = work done (J, N × m, ft × lb_f, N × m × deg or rad, ft × lb_f × deg, or rad)

F = force applied

x = linear distance traveled

T = torque applied

θ = angular distance traveled

The amount of work done by a system must equal the energy put into it minus any friction losses (which may be considered as heat energy gained). [Shortley and Williams 1971, 102–107]

3.1.2 Impulse

Impulse is force applied over a given time period. It has units of force × time for linear systems or force × distance × time

$$I = F \times t \text{ for linear systems} \quad (3.3)$$

$$I = T \times t \text{ for rotary systems} \quad (3.4)$$

where I = impulse imparted to an object or system (N × s, lb_f × s, N × m × s, lb_f × ft × s)

F = force applied

T = torque applied

t = duration at which the force or torque is applied

3.1.3 Momentum

Momentum is mass times velocity for any given object or system. It has units of mass × distance/time for linear systems or mass×distance^2/time for rotary systems

$$M = m \times v \text{ for linear systems} \quad (3.5)$$

$$M = I \times \varpi \text{ for rotary systems} \quad (3.6)$$

where M = momentum of an object or system (kg × m/s, slug × ft/s, kg × m^3 × (deg or rad)/s, lb × ft × s^2 × (deg or rad)/s

m = mass

v = velocity

I = mass moment of inertia

ω = angular velocity

[Shortley and Williams 1971, 129–131]

3.1.4 Energy

Energy can be potential energy or kinetic energy, but in the absence of losses, the sum of all energy into a system has to equal all the energy retained by the system or passed on to another system. When initially looking at a new system, it is often useful to ignore friction. This makes it easier to make sure that conservation of energy occurs. It has the units of force × distance for linear systems or force × distance^2 for rotary systems.

Kinetic energy is given by the following equations:

$$E_k = \frac{1}{2}mv^2 \text{ for linear systems} \tag{3.7}$$

$$E_k = \frac{1}{2}I\omega^2 \text{ for rotaional systems} \tag{3.8}$$

where E_k = kinetic energy (J or N × m, ft × lb_f, N × m × [deg or rad], ft × lb_f × [deg or rad])

m = mass

v = velocity

I = mass moment of inertia

ω = angular velocity

Potential energy can take many forms, but the most common include

$$E_p = \frac{1}{2}kx^2 \text{ for linear systems} \tag{3.9}$$

$$E_p = \frac{1}{2}k\theta^2 \text{ for rotational systems} \tag{3.10}$$

or

$$E_p = mgx \tag{3.11}$$

where E_p = potential energy (Joule or N×m, ft×lb_f, N×m×[deg or rad], ft×lb_f×[deg or rad])

k = spring rate

x = linear displacement

θ = angular displacement

m = mass

g = gravitational constant

[Shortley and Williams 1971, 107–114]

3.1.5 Power

Power is the rate at which work is done. It has units of force × distance/time for linear systems or force×distance²/time

$$P = F \times v \text{ for linear systems} \tag{3.12}$$

$$P = T \times \omega \text{ for rotational systems} \tag{3.13}$$

where P = Power (Watt, Joule/s, N×m/s, ft×lbf/s, Watt×m, Joule×m/s, N×m²/s, ft×lbf×ft/s

F = Force

v = linear velocity

ω = angular velocity

It is useful to remember the relationship between these quantities because they are useful in solving for unknown factors. In the absence of any losses, some relationships include the following or any combination of the following:

- work performed = energy in
- momentum out = impulse imparted
- kinetic energy in = potential energy stored
- power in = work done per unit of time

[Shortley and Williams 1971, 120–122]

3.1.6 Equations of Motion

For linear systems the basic equations of motion are

$$x = x_o + v_o t + \frac{1}{2} a_o t^2 + \frac{1}{6} j t^3 \tag{3.14}$$

where x = final position

x_o = initial position

v_o = initial velocity

a_o = initial acceleration

j = jerk

t = elapsed time

The first derivative of this gives velocity or change in position:

$$\frac{dx}{dt} = v = v_o + a_o t + \frac{1}{2} j t^2 . \tag{3.15}$$

Solving for t yields the time to go from v_o to v with a given a_o and j

$$t = \frac{-a_o \pm \sqrt{a_o^2 - 2j(v_o - v)}}{j} . \tag{3.16}$$

If

$$j = \frac{-a_o}{t} \tag{3.17}$$

then

$$t = \frac{-a_o \pm \sqrt{a_o^2 - 2(-a_o/t)(v_o - v)}}{-a_o/t} \tag{3.18}$$

$$t = \frac{2(v - v_o)}{a_o} . \tag{3.19}$$

The second derivative gives acceleration or the change in velocity:

$$\frac{d^2x}{dt^2} = a = a_o + jt \;. \tag{3.20}$$

The third derivative gives jerk or the change in acceleration:

$$\frac{d^3x}{dt^3} = j \;. \tag{3.21}$$

There is a similar set of equations for rotary systems where the basic equations of motion are

$$\theta = \theta_o + \omega_o t + \frac{1}{2}\alpha_o t^2 + \frac{1}{6}it^3 \tag{3.22}$$

where θ = final position

θ_o = initial position

ω_o = initial velocity

α_o = initial acceleration

ι = jerk

t = elapsed time

The first derivative of this gives velocity or change in position

$$\frac{d\theta}{dt} = \omega_o + \alpha_o t + \frac{1}{2}\iota t^2 \;. \tag{3.23}$$

Solving for t yields the time to go from ω_o to ω with a given ω_o and ι

$$t = \frac{-\alpha_o \pm \sqrt{\alpha_o^2 - 2\iota(\omega_o - \omega)}}{\iota} \;. \tag{3.24}$$

If

$$\iota = \frac{-\alpha_o}{t} \tag{3.25}$$

then

$$t = \frac{-\alpha_o \pm \sqrt{\alpha_o^2 - 2(-\alpha_o/t)(\omega_o - \omega)}}{-\alpha_o/t} \tag{3.26}$$

$$t = \frac{2(\omega - \omega_o)}{\alpha_o} \;. \tag{3.27}$$

The second derivative gives acceleration or the change in velocity:

$$\frac{d^2\theta}{dt^2} = \alpha_o + \iota t \;. \tag{3.28}$$

The third derivative gives jerk or the change in acceleration:

$$\frac{d^3\theta}{dt^3} = \iota \;. \tag{3.29}$$

[Shortley and Williams 1971, 30–46], [Swokowski 1975, 156–160], [Beer and Johnson 1976, 426], [Reswick and Taft 1967, 21–23]

In a high-performance system, it is necessary to operate smoothly and keep from operating near system natural frequencies. This allows it to be operated rapidly using the smallest, most efficient components. If one uses the above equations, a move profile can look something like those in Figure 3.1, where abrupt velocity and acceleration transitions are avoided.

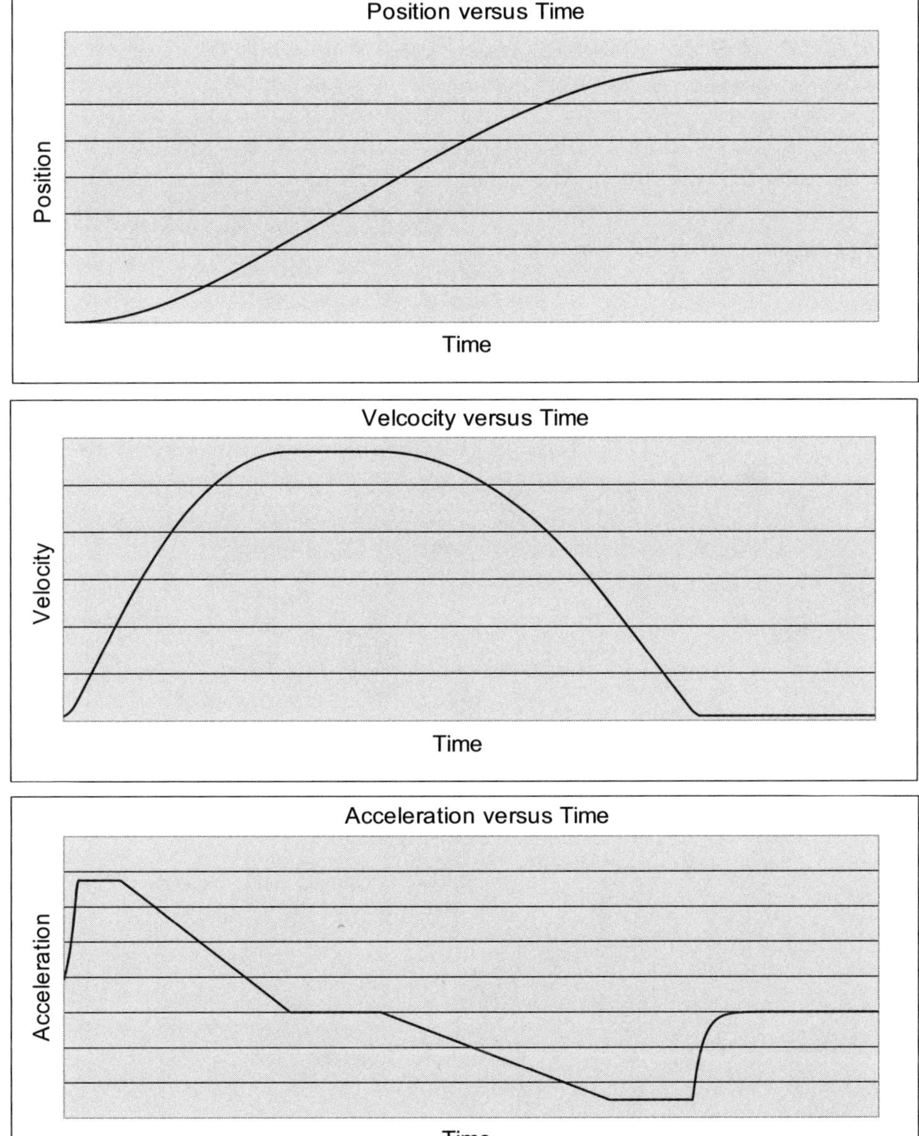

Figure 3.1 Move profiles.

Maintaining smooth operation and stability becomes more important as force, power, velocity, and acceleration increases. The total energy in a system can be summarized by looking at the conservation of energy.

Kinetic Energy E_K – Potential Energy $E_P = 0$

$$\frac{1}{2}mv^2 - \frac{1}{2}kx^2 = 0 \tag{3.30}$$

where m = mass

v = velocity

k = spring rate

x = spring compression

In a dynamic system, the system keeps moving in a given direction while its kinetic energy is greater than the stored potential energy. When E_P is greater than E_K, the system stops and reverses direction and continues to oscillate until all the energy is dissipated. To keep the system moving without oscillation:

$$E_K \geq E_P \tag{3.31}$$

$$\frac{1}{2}mv^2 \geq \frac{1}{2}kx^2 \tag{3.32}$$

$$mv^2 \geq kx^2 . \tag{3.33}$$

Knowing $F = kx$ and $F = ma$, it follows that:

$$x = \frac{ma}{k} \tag{3.34}$$

Using (3.34) with (3.33)

$$mv^2 \geq k\left[\frac{ma}{k}\right]^2 \tag{3.35}$$

$$mv^2 \geq \frac{m^2a^2}{k} \tag{3.36}$$

$$\frac{k}{m} \geq \frac{a^2}{v^2} \tag{3.37}$$

$$\frac{a}{v} \leq \sqrt{\frac{k}{m}} \tag{3.38}$$

or, knowing from (1.1), $\omega_n = \sqrt{\frac{k}{m}}$

$$\frac{a}{v} \leq \omega_n \tag{3.39}$$

where ω_n = natural frequency

If there are losses in the system reducing the potential energy available to reverse the actuator:

$$\frac{1}{2}mv^2 \geq \frac{1}{2}kx^2 \text{ Losses} \tag{3.40}$$

if Losses = $z\%$ of the kinetic energy

$$\frac{1}{2}mv^2 \geq \frac{1}{2}kx^2(1-z) \tag{3.41}$$

$$mv^2 \geq (1-z)kx^2\,. \tag{3.42}$$

Knowing $F = kx$ and $F = ma$

$$x = \frac{ma}{k} \tag{3.43}$$

$$mv^2 \geq (1-z)k\left[\frac{ma}{k}\right]^2 \tag{3.44}$$

$$mv^2 \geq \frac{(1-z)m^2a^2}{k} \tag{3.45}$$

$$\frac{k}{m} \geq \frac{(1-z)a^2}{v^2} \tag{3.46}$$

$$\frac{a}{v} \leq \sqrt{\frac{k}{m(1-z)}} \tag{3.47}$$

or

$$\frac{a}{v} \leq \frac{\omega_n}{\sqrt{1-z}}\,. \tag{3.48}$$

If one takes a conservative approach, losses are ignored, so the question is how much below ω_n should $\frac{a}{v}$ be? If the design margin is 4, then the solution is

$$\frac{1}{2}mv^2 \geq 4 \times \frac{1}{2}kx^2 \tag{3.49}$$

$$\frac{mv^2}{2} \geq 2 \times kx^2 \tag{3.50}$$

$$mv^2 \geq 4 \times k\left[\frac{ma}{k}\right]^2 \tag{3.51}$$

$$mv^2 \geq \frac{4 \times m^2a^2}{k} \tag{3.52}$$

$$\frac{k}{4 \times m} \geq \frac{a^2}{v^2} \tag{3.53}$$

$$\frac{a}{v} \le .5\sqrt{\frac{k}{m}} \tag{3.54}$$

or

$$\frac{a}{v} \le .5\omega_n\,. \tag{3.55}$$

Generalizing this so that $C = \frac{1}{\sqrt{\text{Design Margin}}}$

$$\frac{a}{v} \le C\omega_n \text{ or } \frac{dv}{dx} \le C\omega_n\,. \tag{3.56}$$

One can see that maintaining this relationship becomes more of an issue if acceleration is maintained at low velocities. Relationships at the transition between constant acceleration and a = $Cv\omega_n$ can be determined as follows:

$$x = x_o + v_o t + \frac{1}{2}a_o t^2 \tag{3.57}$$

$$v = v_o + a_o t \tag{3.58}$$

$$t = \frac{-v_o \pm \sqrt{v_o^2 - 2a_o(x_o - x)}}{a_o} \tag{3.59}$$

$$t = \frac{v - v_o}{a_o} \tag{3.60}$$

$$\frac{v - v_o}{a} = \frac{-v_o \pm \sqrt{v_o^2 - 2a_o(x_o - x)}}{a} \tag{3.61}$$

$$v = \pm\sqrt{v_o^2 - 2a_o(x_o - x)}\,. \tag{3.62}$$

As $v > 0$

$$v = \sqrt{v_o^2 - 2a_o(x_o - x)} \tag{3.63}$$

$$a = \frac{v^2 - v_o^2}{2(x - x_o)} \tag{3.64}$$

and a look at the change in velocity with respect to position results in

$$\frac{dv}{dx} = \frac{1}{2}\left[v_o^2 - 2a(x_o - x)\right]^{-\frac{1}{2}}(2a) \tag{3.65}$$

$$\frac{dv}{dx} = \frac{a}{\sqrt{v_o^2 - 2a(x_o - x)}}\,. \tag{3.66}$$

Knowing $\frac{dv}{dx} = C\omega_n$

$$C\omega_n = \frac{a}{\sqrt{v_o^2 - 2a(x_o - x)}} \tag{3.67}$$

$$x = x_o + \frac{a}{2(C\omega_n)^2} - \frac{v_o^2}{2a} \tag{3.68}$$

and also knowing

$$\frac{dv}{dx} = C\omega_n \tag{3.69}$$

$$\frac{dv}{dt} = a \tag{3.70}$$

$$dt = \frac{C\omega_n}{a}dx \tag{3.71}$$

$$\int_{t_o}^{t} t = \int_{x_o}^{x} \frac{C\omega_n}{a}dx \tag{3.72}$$

$$t = \frac{C\omega_n}{a}(x - x_o) + t_o \tag{3.73}$$

however, t goes to infinity as a goes to 0

$$a = \frac{C\omega_n}{t - t_o}(x - x_o) \tag{3.74}$$

and

$$\frac{dx}{dt} = v \tag{3.75}$$

$$\frac{dv}{dx} = C\omega_n \tag{3.76}$$

$$dt = \frac{1}{(C\omega_n)v}dv \tag{3.77}$$

$$\int_{t_o}^{t} dt = \frac{1}{C\omega_n}\int_{v_o}^{v} \frac{1}{v}dv \tag{3.78}$$

$$t = \frac{1}{C\omega_n}\ln\left(\frac{v}{v_o}\right) + t_o \tag{3.79}$$

again, t goes to infinity as v_o goes to 0

$$e^{C\omega_n(t-t_o)} = \frac{v}{v_o} \tag{3.80}$$

$$v = v_o e^{C\omega_n(t-t_o)} \tag{3.81}$$

$$x = \int_{t_o}^{t} v_o e^{C\omega_n(t)}dt \tag{3.82}$$

$$x = \frac{v_o}{C\omega_n} e^{C\omega_n(t-t_o)} + \text{Constant} . \tag{3.83}$$

At the start, $t = 0$, $x = 0$

$$0 = \frac{v_o}{C\omega_n} e^{C\omega_n(0)} + \text{Constant} \tag{3.84}$$

$$-\frac{v_o}{C\omega_n} = \text{Constant} \tag{3.85}$$

$$x = \frac{v_o}{C\omega_n} e^{C\omega_n(t-t_o)} - \frac{v_o}{C\omega_n} \tag{3.86}$$

$$x = \frac{v_o}{C\omega_n}\left[e^{C\omega_n(t-t_o)} - 1\right] \tag{3.87}$$

$$a = \frac{dv}{dt} \tag{3.88}$$

$$a = v_o C\omega_n e^{C\omega_n(t-t_o)} \tag{3.89}$$

$$\frac{a}{V_o C\omega_n} = e^{C\omega_n(t-t_o)} \tag{3.90}$$

$$\ln\left(\frac{a}{v_o C\omega_n}\right) = C\omega_n(t - t_o) \tag{3.91}$$

$$t = \frac{1}{C\omega_n}\ln\left(\frac{a}{v_o C\omega_n}\right) + t_o \tag{3.92}$$

but there is no valid solution for $\dfrac{a}{v_o C\omega_n} < 0$.

Similarly

$$dv = C\omega_n dx \tag{3.93}$$

$$\int_{v_o}^{v} v = C\omega_n \int_{x_o}^{x} dx \tag{3.94}$$

$$x = \frac{v - v_o}{C\omega_n} + x_o . \tag{3.95}$$

While this is manageable, some may wish to simplify the equations to approximate $\frac{a}{v} \leq C\omega_n$ or $\frac{dv}{dx} \leq C\omega_n$ with jerk *j*. From the preceding information, one can see the desire to manage velocity changes within the limits of the system natural frequency. The change in velocity with respect to position should maintain a value less than or equal to system natural frequency with the appropriate design margin. As the system decelerates, acceleration/deceleration needs to get smaller according to the relationship, $a = C\omega_n v$. A system that follows this relationship has stable performance at the ends of the cycle. As the system approaches or leaves the constant velocity portion of the motion profile, strictly following $a = C\omega_n v$ creates an abrupt transition between acceleration/deceleration and the maximum velocity portion of the motion profile. Studying $a = C\omega_n v$, one can see that with $a = C\omega_n v$ as a constant, acceleration a increases or decreases with velocity v. Therefore, the limiting change in acceleration occurs at the minimum velocity value for each transition. At the ends, this is at the initial or final velocity. As the system approaches maximum velocity, the limiting velocity occurs at the acceleration starts decreasing (j is negative). As the system leaves maximum velocity, it occurs as the system reaches its constant deceleration value. Because the transitions to/from maximum velocity need to take place at constant jerk j, we need to understand the relationship between natural frequency and jerk at the transition velocity.

Knowing

$$x = x_o + v_o t + \frac{1}{2} a_o t^2 + \frac{1}{6} j t^3 \tag{3.96}$$

$$\frac{dx}{dt} = v = v_o + a_o t + \frac{1}{2} j t^2 \tag{3.97}$$

$$\frac{d^2 x}{dt^2} = a = a_o + jt \tag{3.98}$$

$$\frac{d^3 x}{dt^3} = j \tag{3.99}$$

and $\frac{a}{v} \leq C\omega_n$, we can get

$$\frac{a_o + jt}{v_o + at + \frac{1}{2} j t^2} = C\omega_n \tag{3.100}$$

or

$$j = \frac{C\omega_n v_o + C\omega_n a_o t + a_a}{t - \frac{C\omega_n t^2}{2}}. \tag{3.101}$$

We also know

$$v = v_o e^{C\omega_n (t - t_o)} \tag{3.81}$$

and

$$a = v_o C\omega_n e^{C\omega_n (t - t_o)} \tag{3.89}$$

Therefore

$$\frac{da}{dt} = j = (C\omega_n)^2 v_o. \tag{3.102}$$

So it follows that:

$$j = C\omega_n a \tag{3.103}$$

knowing a_o, v_o, C, and ω_n, one can solve for j at the t of their choice where the acceleration will be a and velocity v.

3.1.6.1 Summary

3.1.6.1.1 Velocity

Figure 3.2 contains two profiles: a simple velocity/time profile that just looks at constant velocities and constant accelerations in a move between two points, A and B. For this profile, the velocity equation of motion is [see (3.58) and (3.97)]

$$v = v_o + at.$$

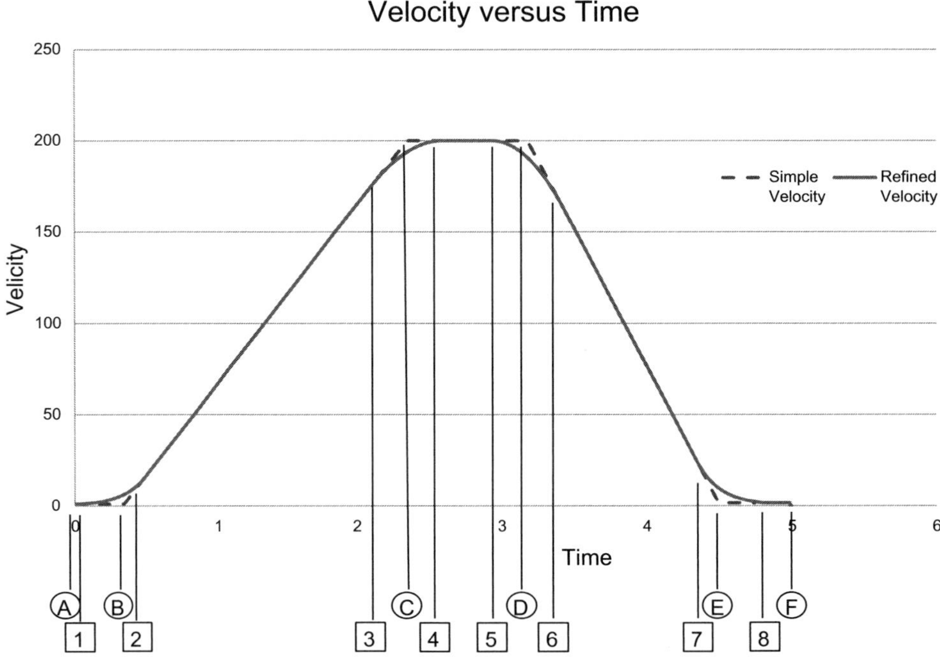

Figure 3.2 Velocity/time profile.

The refined velocity profile allows for smooth transitions between periods of constant velocity and constant acceleration. The velocity equations of motion in the refined profile (remember to use the appropriate "sign" for the direction or condition) are as follows:

1. *From A to 1:*

$$V = \text{initial velocity } (v_o)$$

2. *From 1 to 2 and 7 to 8:*

$$v = v_{1,7} e^{C\omega_n (t - t_{1,7})} \tag{3.81}$$

3. *From 2 to 3 and 6 to 7:* [see (3.57) and (3.96)]

$$v = v_{2,6} + a_{2,6}\left(t - t_{2,6}\right)$$

4. *From 3 to 4 and 5 to 6:* [see (3.96)]

$$v = v_{3,5} + a_{3,5}t_{3,5} + \frac{1}{2}j_{3,6}\left(t - t_{3,5}\right)^2$$

where $j = C\omega_n$ for ω_n at the lowest velocity (i.e., positions 3 and 6)

5. *From 4 to 5 (Maximum Velocity):* [see (3.57) and (3.96)]

$$v = v_4.$$

6. *From 8 to F:*

$$V = \text{final velocity } (v_f).$$

3.1.6.1.2 Acceleration

Figure 3.3 contains two profiles: a simple acceleration/time profile that just looks at constant velocities and constant accelerations in a move between two points, A and B. For this profile, the acceleration equation of motion is [see (3.98)]

$$a = a_{B,D}.$$

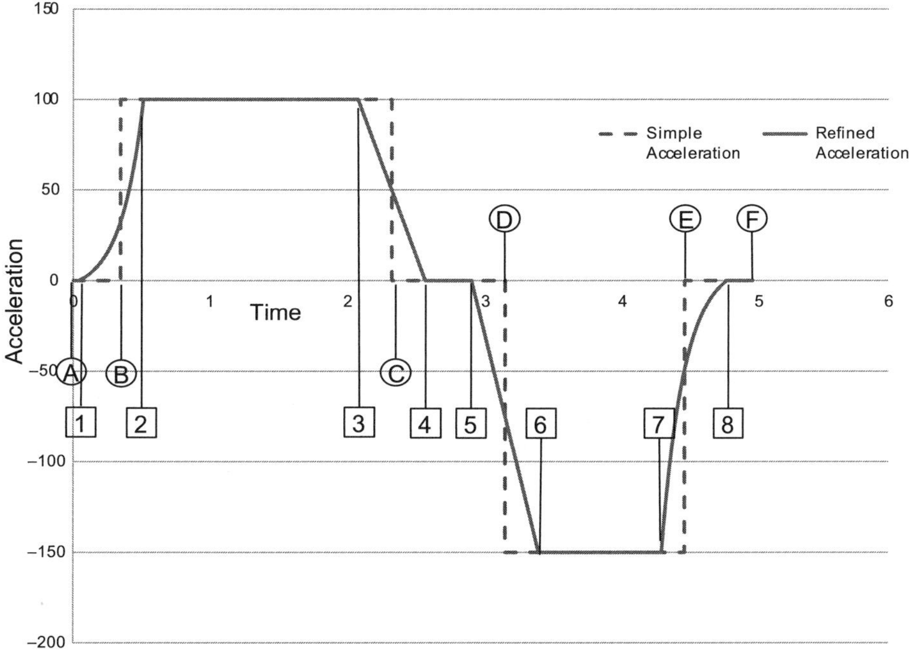

Figure 3.3 Acceleration/time profile.

The refined acceleration profile allows for smooth transitions between periods of constant velocity and constant acceleration. From 1 to 2, and 7 to 8, a constant change of velocity with respect to position is used to keep within system natural frequency limits. From 3 to 4 and 5 to 6, a constant jerk value is used where jerk is determined from the lowest change in velocity with respect to position. This value is at the transition velocity between constant acceleration and constant velocity. As velocity continues to increase, utilizing the same transition as between 1/2 and 7/8 does not create a smooth transition to a maximum constant velocity. The acceleration equations of motion in the refined profile (remember to use the appropriate "sign" for the direction or condition) are as follows:

1. *From A to 1:*

$$a = 0.$$

2. *From 1 to 2 and 7 to 8:* [see (3.89)]

$$a = v_1 C\omega_n e^{C\omega_n (t-t_o)}$$

$$a = v_8 C\omega_n e^{C\omega_n (t-t_o)}.$$

3. *From 2 to 3 and 6 to 7:*

$$a = a_{2,6}.$$

4. *From 3 to 4 and 5 to 6:* [see (3.96)]

$$a = a_{3,5}\left(t - t_{3,5}\right) + \frac{1}{2} j_{3,6}\left(t - t_{3,5}\right)^2$$

where $j = C\omega_n$ for ω_n at the lowest velocity (i.e., positions 3 and 6).

5. *From 4 to 5 (Maximum Velocity):* [see (3.57) and (3.96)]

$$a = 0.$$

6. *From 8 to F:*

$$a = 0.$$

3.1.6.1.3 Position

Figure 3.4 contains two profiles: a simple position/time profile that just looks at constant velocities and constant accelerations in a move between two points, A and B. For this profile, the position equation of motion is [see (3.57) and (3.96)]

$$x = x_o + v_o t + \frac{1}{2} a_o t^2.$$

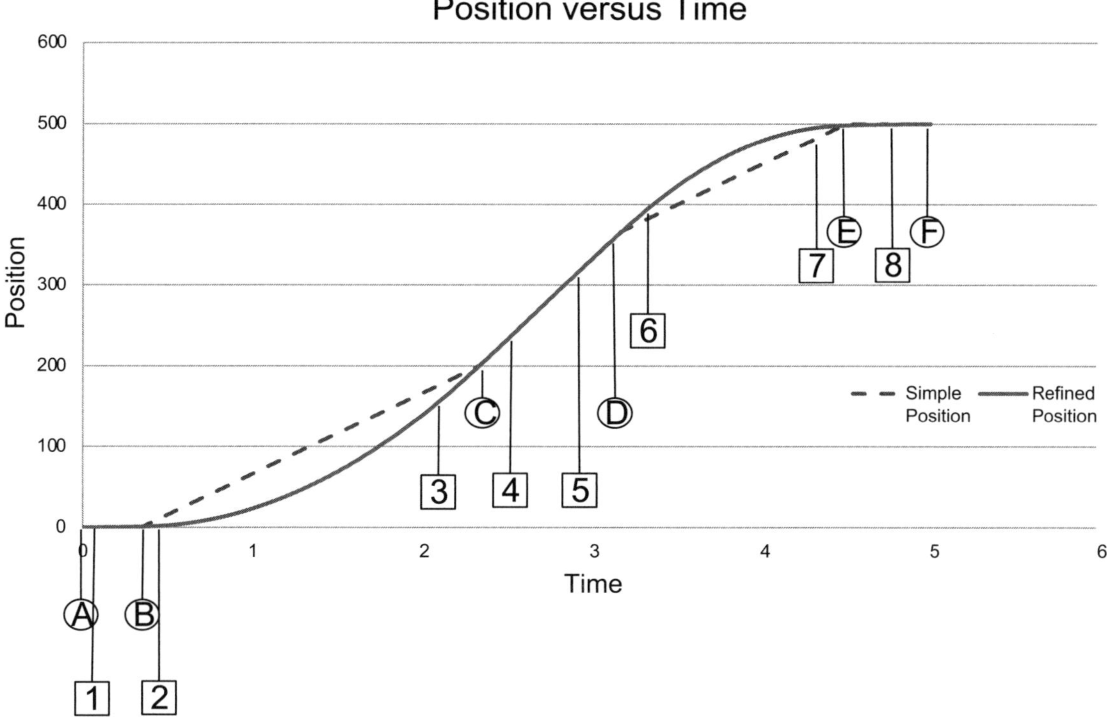

Figure 3.4 Position/time profile.

The refined position profile allows for smooth transitions between periods of constant velocity and constant acceleration. The position equations in the refined profile (remember to use the appropriate "sign" for the direction or condition) are as follows:

1. *From A to 1:* [see (3.57)]

$$x = x_o + v_o t$$

2. *From 1 to 2 and 7 to 8:* [see (3.87)]

$$x = \frac{v_{1,7}}{C\omega_n}\left[e^{C\omega_n(t-t_{1,7})} - 1\right]$$

3. *From 2 to 3 and 6 to 7:* [see (3.57)]

$$x = x_{2,6} + v_{2,6}\left(t - t_{2,6}\right) + \frac{1}{2}a_{2,6}\left(t - t_{2,6}\right)^2$$

4. *From 3 to 4 and 5 to 6:* [see (3.96)]

$$x = x_{3,5} + v_{3,5}\left(t - t_{3,5}\right) + \frac{1}{2}a_{3,5}t^2 + \frac{1}{6}j_{3,5}\left(t - t_{3,5}\right)^3$$

where $j = C\omega_n$ for ω_n at the lowest velocity (i.e., positions 3 and 6).

5. *From 4 to 5 (Maximum Velocity):* [see (3.57)]

$$x = x_4 + v_4(t - t_4).$$

6. *From 8 to F:* [see (3.57)]

$$x = x_8 + v_8(t - t_8).$$

3.2 Loads

Often the design requirements as given do not define the actuator but rather what the load needs to do. Therefore, the engineer's first task is to develop an initial concept of how the product (machine) is going accomplish its task. Using information from the concept, the type of actuator and loads on the actuator can be determined. Sometimes this is straight forward and easy. Other times, translating load motion and loads into actuator motion and loads requires system modeling and translating the equations of motion so that they can be used for the application. Actuator loads flow from Newton's second law of motion:

1. *For linear systems:*

$$F = ma \tag{3.104}$$

where F = actuator force required, m = actuator load mass, and a = acceleration.

2. *For rotary systems:*

$$T = I\alpha \tag{3.105}$$

where T = actuator torque required, I = actuator load inertia, and α = acceleration. [Meriam 1975, 6]

It is important to remember that the mass or inertia in these equations must be the total mass or inertia seen by the actuator, including all structures and mechanisms. If one assumes that there are no losses in the system, then the power transmitted through the system must be constant or $P_1 = P_2$.

Because

$$P = F \times v \text{ for linear systems} \tag{3.106}$$

$$P = T \times \varpi \text{ for rotational systems} \tag{3.107}$$

and

$$F = ma \tag{3.108}$$

$$T = I\alpha \tag{3.109}$$

then

$$m_1 a_1 v_1 = m_2 a_2 v_2. \tag{3.110}$$

If there is some speed or gear ratio, and R is between 1 and 2, then

$$a_1 = R \times a_1 \tag{3.111}$$

$$v_1 = R \times v_1 \tag{3.112}$$

$$m_1 \times a_1 \times v_1 = m_2 \times R \times a_1 \times R \times v_1 \tag{3.113}$$

or the equivalent mass, at the actuator or m_1 location is

$$m_1 = R^2 \times m_2 \tag{3.114}$$

The same is true for rotary systems where

$$I_1 = R^2 \times I_2. \tag{3.115}$$

These relationships can be used throughout the drive system to determine structural loads and loads seen by the actuator. These loads can be used to make sure structural and actuator components are properly sized for the loads experienced. As discussed earlier, for high-performance actuators, it is also important to evaluate the stiffness of the actuator and associated structural components.

A sample linear actuator powered by a motor is shown in Figure 3.5:

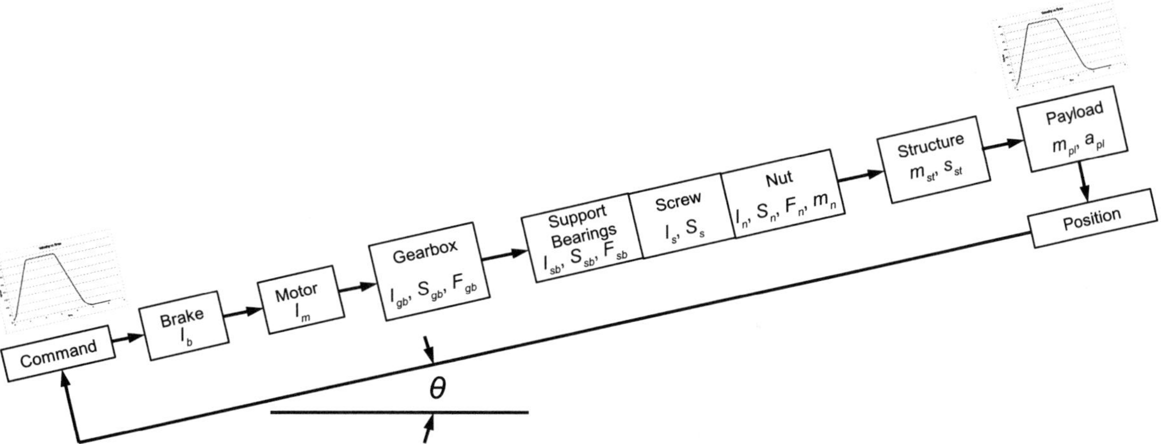

Figure 3.5 Linear actuator powered by a motor.

for θ = payload angle from horizontal

a_L = load acceleration

g = acceleration due to gravity

μ_L = load coefficient of friction

m = load mass

I_{Tm} = total inertia at the motor

I_b = brake inertia

I_m = motor inertia

I_G = gear box inertia

R_G = gear box ratio

I_s = screw inertia

I_n = nut inertia

F = load force requirement

F_R = force required to raise payload

F_L = force required to lower payload

α_m = motor acceleration

lead = screw lead

eff_s = screw efficiency

eff_G = gear box efficiency

T_{PL} = screw/nut preload torque

S_m = stiffness at the motor

S_G = gear box stiffness

S_{GBon} = stiffness between the gear box and load

S_B = support bearing stiffness (linear)

S_s = screw stiffness (linear)

S_n = nut stiffness (linear)

S_{Str} = structure stiffness (linear)

Torque to exert a force F with a screw is given by

$$T = \frac{F(\text{lead})}{2\pi(\text{eff}_s)} \tag{3.116}$$

Force required to raise payload is given by

$$F_R = m\left(a_L + g\sin(\theta)\right) + \mu_L mg\cos(\theta) \tag{3.117}$$

Force required to lower payload is given by

$$F_L = m\left(a_L + g\sin(\theta)\right) + \mu_L mg\cos(\theta)\,. \tag{3.118}$$

[Spotts 1978, 249–254]

Inertia at the motor is given by

$$I_{Tm} = I_b + I_m + I_G + \frac{1}{R_G^2} I_s + \frac{1}{R_G^2} I_n\,. \tag{3.119}$$

Motor torque is given by

$$T_m = I_{Tm}\alpha_m + \frac{1}{R_G\text{eff}_G}\left[\frac{F(\text{lead})}{2\pi(\text{eff}_s)}\right] + \frac{1}{R_G\text{eff}_G}T_{\text{PL}}. \tag{3.120}$$

Linear stiffness translated to rotary stiffness is given by

$$S_{\text{GBon}} = \frac{\dfrac{(\text{lead})^2}{\left[\dfrac{1}{S_B}+\dfrac{1}{S_s}+\dfrac{1}{S_n}+\dfrac{1}{S_{\text{Str}}}\right]}}{2\pi(\text{eff}_s)} \tag{3.121}$$

Stiffness of the motor is given by

$$\frac{1}{S_m} = \frac{1}{S_G} + \frac{R_G}{S_{\text{GBon}}}. \tag{3.122}$$

Stiffness of the motor is then given by

$$S_m = \frac{1}{\left[\dfrac{1}{S_G} + \dfrac{R_G}{S_{\text{GBon}}}\right]}. \tag{3.123}$$

3.3 Component Stiffness

As we saw earlier in Section 3.1.6, stiffness of the actuator system plays an important role in the performance level of an actuator. The ideal actuator system would be infinitely stiff, but real-world design considerations such as cost, weight, and space tend to drive designs toward minimal stiffness. The engineer needs to consider the stiffness of the entire system, including actuator components, supporting structure, and even the load. One can break each contributor to system stiffness into basic elements, determine the stiffness of the element, and then combine them to get system stiffness. While there can be more complicated arrangements, the fundamental stiffness characteristics are either linear or rotational.

3.3.1 Linear Stiffness

In general, stiffness is characterized by the spring rate k of a component or system. For linear systems k is in terms of force per unit of displacement, often seen in the following equation for linear spring force

$$F = kx \tag{3.124}$$

where F = force, k = spring rate, and x = displacement [Reswick and Taft 1967, 11]

For many components under direct linear loading such as those shown in Figure 3.6, there is a linear relationship between deflection and stress of a material that follows Hooke's law in the elastic range of a material. This is often as shown in

$$E = \frac{\sigma}{\varepsilon} \tag{3.125}$$

where E = Modulus of elasticity, σ = stress, and ε = strain. [Byars and Snyder 1975, 69, 77]

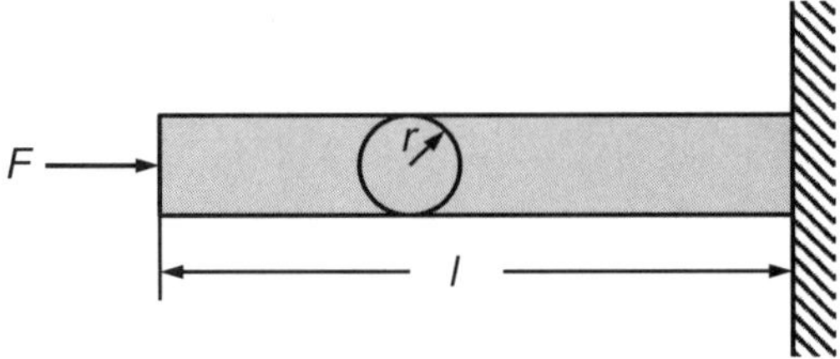

Figure 3.6 Bar under direct linear load, where F = applied force, l = bar length, r = bar radius, and A = bar area (πr^2).

Stress (σ) is the force divided by the area that the force is applied to and strain (ε) is the change in length divided by the original length:

$$\sigma = \frac{F}{A} \tag{3.126}$$

where σ = stress, F = force, and A = area.

Strain for this bar can be represented as follows in:

$$\varepsilon = \frac{\Delta l}{l} \tag{3.127}$$

where ε = strain, Δl = change in bar length, and l = original bar length.

Combining (3.125)–(3.127), we get the spring rate for the directly loaded bar:

$$k = \frac{F}{\Delta l} = \frac{E \times A}{l}. \tag{3.128}$$

3.3.2 Rotary Stiffness

For systems under torque load, stiffness is characterized by the torsional spring rate k of a component or system. For rotary systems, k is in terms of torque per unit of rotary displacement, often represented by the following equation of torsional spring force:

$$T = k\varphi \tag{3.129}$$

where T = torque, k = torsional spring rate, and φ = displacement.

For many components under direct torsion loading such as shown in Figure 3.7, there is a relationship between deflection and shear stress of a material that follows Hooke's law in elastic range of a material. This is can be represented by

$$G = \frac{\tau}{\gamma} \tag{3.130}$$

where G = shear modulus of elasticity, τ = shear stress, and γ = shear strain. [Byers and Snyder 1975, 70]

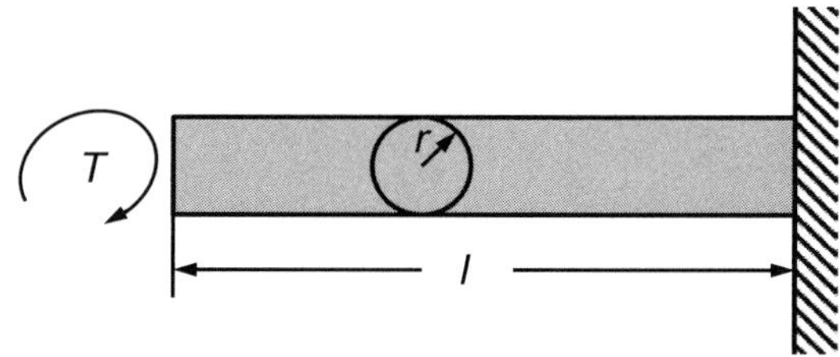

Figure 3.7 Bar under direct torsion load, where T = applied force, l = bar length, r = bar radius, and A = bar area (πr^2).

For a round member, we know that shear stress (τ) is the torque times radius divided by the polar moment of inertia (J) and shear strain (γ) is the change in bar angle times the radius divided by the bar length:

$$\tau = \frac{T \times r}{J} \tag{3.131}$$

where τ = shear stress, T = applied torque, and J = polar moment of inertia (in this example, $J = \frac{\pi r^4}{2}$).

Similar to the strain for the directly loaded bar of Figure 3.7, shear strain can be represented by

$$\gamma = \frac{r \times \Delta\varphi}{l} \tag{3.132}$$

where γ = shear strain, r = bar radius, $\Delta\varphi$ = change is bar angle (radians), and l = original bar length. [Spotts 1978, 132–135]

Combining (3.130)–(3.132), we can get the torsional spring rate for the bar under direct torsion load in Figure 3.7:

$$k = \frac{T}{\Delta\varphi} = \frac{G \times J}{l}. \tag{3.133}$$

In reality, an actuator system is made up of many springs combined in parallel and series as shown in Figure 3.8.

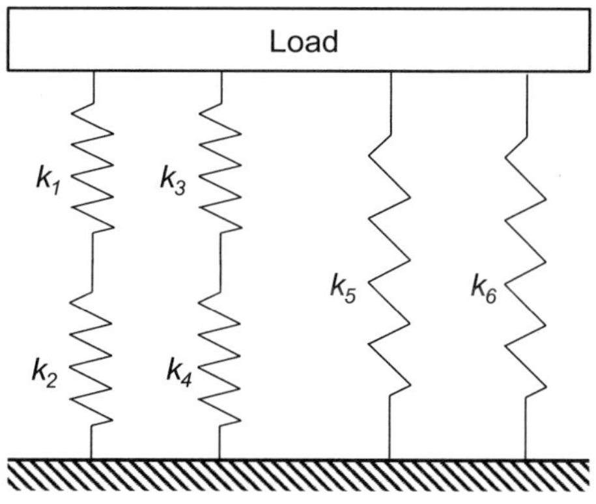

Figure 3.8 Generic actuator system.

From basic mechanical vibrations, we know that springs can be combined to create an equivalent spring rate for the system. The equivalent spring rate for springs in parallel and the equivalent spring rate for springs in series are shown, respectively, in [Harris and Crede 1976, 2-2 to 2-4, 2-16 to 2-18, 10-17 to 10-19]

$$k_{\text{eq parallel}} = k_1 + k_2 + \cdots \quad (3.134)$$

$$k_{\text{eq series}} = \frac{1}{1/k_1} + \frac{1}{1/k_2} + \cdots \quad (3.135)$$

The equivalent spring rate for the complete system shown in Figure 3.8 is shown in

$$k_{\text{eq}} = \frac{1}{1/k_1} + \frac{1}{1/k_2} + \frac{1}{1/k_3} + \frac{1}{1/k_4} + k_5 + k_6 \quad (3.136)$$

Stiffness was discussed in the equations of motion, where we saw how it influences the change in acceleration a design can handle without exciting system natural frequencies $(f_n = \sqrt{k/m})$. As one of the goals of high-performance actuators is usually to operate rapidly, a stiff system has the inherent potential to operate faster than a less stiff system. This book does not address the specific process for all stiffness calculations because the engineer is expected to have gained that knowledge from standard statics, dynamics, engineering mechanics, strength of materials, and other engineering design and analysis courses. However, as this is so important and is an often overlooked area of design, a few areas of consideration are covered in the details of various actuator components. For example, power screw stiffness is covered in Section 5.6.4, hydraulic cylinders in Section 5.6.1.1, and hydraulic motors in Section 5.4.1.5.

3.3.3 Materials

Materials selected for a design can have a significant impact on its stiffness. Tables 3.4 and 3.5 capture some material properties of interest in designing actuators. They do not capture all engineering materials but rather illustrate how various materials compare with each other in stiffness and mass along with some thermal properties. It is important to note that the natural frequency is a function of both stiffness and mass, so an understanding of the relationship between the two helps the engineer design the best actuator for a given application. It is the combination of stiffness and mass that matters! The design engineer should also realize that the elastic modulus and bulk modulus of materials change with temperature, going down as the temperature goes up. For "normal" operating temperature ranges, this may not be a significant factor for metallic components and some structural composites but could become significant for lower modulus materials, especially at high temperatures. Some general properties of common metals are shown in Table 3.4. For designs under significant weight or performance pressure, it can be helpful to calculate the specific stiffness as done in Table 3.5.

Table 3.4 Material properties

Material	Modulus E GPa	Modulus E ksi	Shear Modulus G GPa	Shear Modulus G ksi	Bulk Modulus K $K = E/[3(1-2v)]$ GPa	Bulk Modulus K ksi	Poison's Ratio v	Density γ g/cc	Density γ lb/in^3
Aluminum	69	10,000	26.2	3,800	67.6	9,804	0.33	2.77	0.10
Titanium	114.0	16,500	42.7	6,200			0.33	4.70	0.17
Steel	207.0	30,000	79.0	11,500	172.5	25,000	0.30	7.83	0.283

Material property representative and actual values may be different.

Table 3.5 Material specific stiffness comparison

Material	Specific Stiffness (GPa-cc/g) $\left(\frac{E}{\gamma}\right)$	Specific Frequency (GPa-cc/g)$^{.5}$ $\left(\sqrt{\left(\frac{E}{\gamma}\right)}\right)$
Aluminum	24.91	4.99
Titanium	24.26	4.92
Steel	26.44	5.14

Material property representative and actual values may be different.

During requirements analysis, it is important to develop an initial concept based on material properties that the engineer can reasonably expect to have in the final configuration. At this stage, the project does not usually want to expend the effort it takes to establish exact material properties, especially when they may change as the overall design matures. However, as with all designs, the engineer should get final material properties from the supplier or manufacturer when developing the final configuration. Good design and documentation practice ties the material specified for a component to a well-established source such as MIL-HDBK-5/MMPDS or ASTM standards whenever possible.

3.4 Component Strength

One of the most important contributions the engineer makes to project and society in general is ensuring that designs have adequate strength to perform all their functions and have some margin for operator error, manufacturing inaccuracies, and degradation over life. Through careful design and analysis, the engineer also makes sure that the product is not significantly over designed. Overdesign can lead to wasted material, excess weight, and greater energy requirements than necessary. The initial design generated to support requirements analysis should have an accompanying preliminary strength analysis so that there is confidence that the product can be built to satisfy the final requirement set. Initial loading can come from the equations of motion generated for the system and resulting loads. Consideration should also be made for other loads such as transportation, environment, and operating through the entire load range as any of these conditions may require greater component strength for operating conditions that are not moving the largest operating load. Techniques for performing strength calculations should be gained from standard engineering design and analysis courses. The various types of strength that the design engineer might need to consider for any give design include yield strength/elastic analysis, ultimate strength/plastic analysis, and fatigue strength. When deciding on which strength criteria and analysis approach should be used, the engineer should consider the type of material and material failure mode, load frequency, and required life of the component or system.

As the design matures, loads from the actuator's environment (mobility, shock, vibration, snow, ice, and temperature differential) and operation (acceleration and deceleration of the load) as defined in environmental requirements and determined through performance analysis and/or testing can be applied to the structure to evaluate structural strength. When the structure and components are adjusted to achieve strength, stiffness, and life goals, performance calculations and modeling may also need to be adjusted to account for the new structural properties. This design iteration process continues until all key requirements are satisfied. As part of this process, the engineer should be conscience of how vibration loads may affect the fatigue life or how loads may excite structural modes of the system. Each load case needs to have a load profile with duty cycle defined so it can be evaluated to the relevant material strength property. Single events may want to be evaluated against the material's ultimate or yield strength, depending on the failure mode. Dynamic and vibration loads may need to be evaluated against the material's stress-life characteristics. The following sections address some of the strength criteria and associated analysis approaches in more detail.

3.4.1 Yield Strength and Elastic Analysis

When one looks at designing a component or system, they generally try to stay in the linear range of a material. Most engineering materials exhibit a linear relationship between stress and strain up to the material's yield strength as defined by Hooke's law (3.125). Utilizing the yield stress as the limit for a design can provide some measure of safety as a component does not immediately break or fail when its material is loaded to or beyond the yield stress. Most finite-element analysis and hand calculations also assume that the material behaves linearly, that is, it is loaded below the yield stress. At any given location in a part, the stress may be due to bending, direct loading, or shear. When determining the strength of a component, it is important to understand the type of stress induced at the various locations and how the different types of stress should be combined.

Bending stress is induced in a part when the applied load causes a moment to be generated. The maximum bending stress in a component is dependent on the bending moment and the area moment of inertia at that location. For the simply supported beam shown in Figure 3.9, the maximum bending stress is at the center of the span. An analysis or hand calculation needs to be done to determine the moment and resulting bending stress for all critical components. The resulting stress field in the beam is shown in Figure 3.10. The bending stress at any point in this field can be calculated from (3.137). The bending moment and shear can be shown in a shear–moment diagram such as Figure 3.9. The reaction forces (*R*) can be calculated for the load (*P*) by summing moments at one of the supports. These reaction forces can then be used to determine the shear by applying them and any other loads along the beam. The moment at any point in the beam can be calculated by multiplying distance between the point of interest and reaction force and/or load(s) and summing them. In either case, the engineer needs to make sure that the direction (sign) of the reaction force or load is applied correctly:

$$\sigma_b = \frac{Mc}{I} \tag{3.137}$$

where σ_b = bending stress, M = bending moment, and c = distance from the neutral axis. [Spotts 1978, 13]

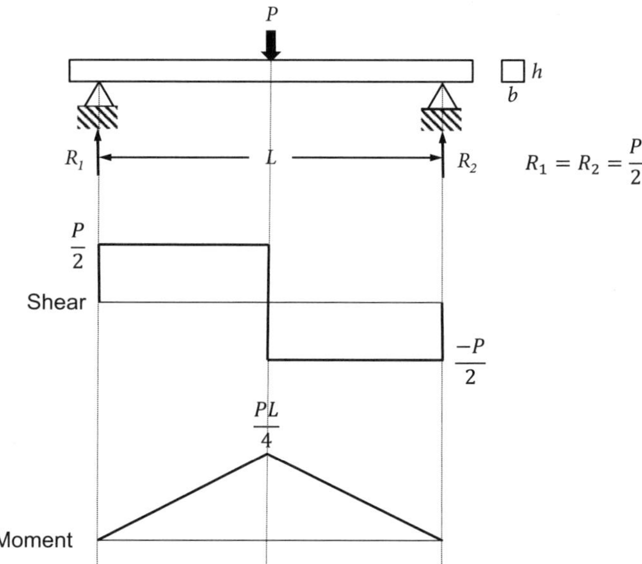

Figure 3.9 Simply supported beam.

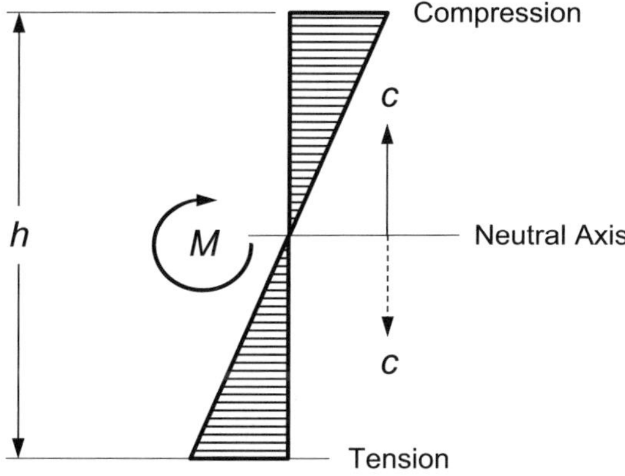

Figure 3.10 Bending moment stress field.

The area moment of inertia is dependent on the geometry of the structural member and needs to be evaluated at the location where the bending stress calculation is made. The general formulas for the area moment of inertia about the three mutually perpendicular axes, x, y, and z, are shown in

$$I_x = \int y^2\,dA$$
$$I_y = \int x^2\,dA \qquad (3.138)$$
$$I_z = \int r^2\,dA$$

For the rectangular section shown in Figure 3.11, these equations along the x-axis become

$$I_x = \int_{-h/2}^{h/2} y^2 b\,dy$$
$$I_x = \left.\frac{by^3}{3}\right|_{-h/2}^{h/2} \qquad (3.139)$$
$$I_x = \frac{bh^3}{12}$$

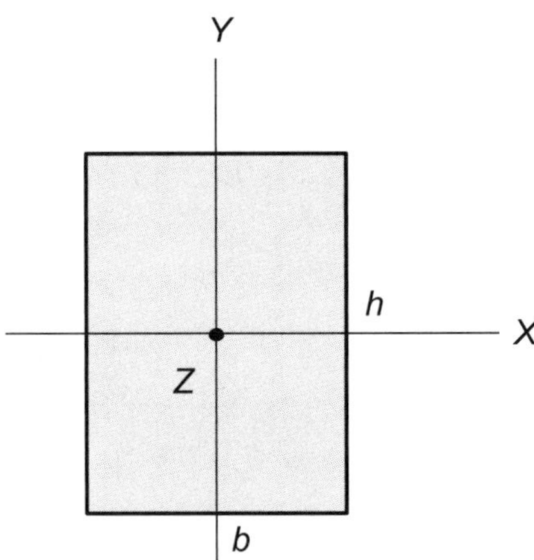

Figure 3.11 Rectangular section.

Likewise, the area moment of inertia of a rectangle about the y-axis is given by

$$I_y = \frac{hb^3}{12} \qquad (3.140)$$

and because $r^2 = x^2 + y^2$, $I_z = I_x + I_y$, and the area moment of inertia of a rectangle about the z-axis is given by

$$I_z = \frac{bh^3}{12} + \frac{hb^3}{12}$$
$$I_z = \frac{bh}{12} + \left(h^2 + b^2\right) \qquad (3.141)$$

If the area moment of inertia of interest is not about the neutral axis of the area, it can be calculated using the parallel axis theorem. In this case, shown is the area moment of inertia along the *a*-axis for the location of interest is a combination of the area moment of inertia about the neutral axis plus the area times the distance of the area neutral axis from the axis of interest squared. For a rectangular area, this is shown in Figure 3.12:

$$I_a = \frac{bh^3}{12} + bh(a)^2 \tag{3.142}$$

[Meriam 1975, 319–323]

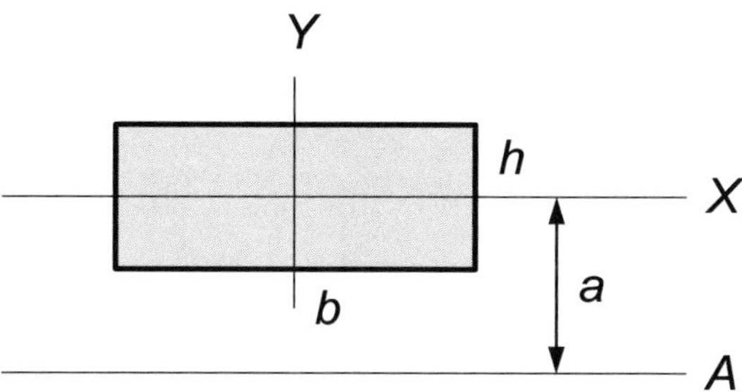

Figure 3.12 Parallel axis theorem.

Through the principle of superposition (adding and subtracting the area moment of inertia combinations) and parallel axis theorem, one can easily determine the area moment of inertia of many common shapes. Figure 3.13 summarizes the area moment of inertia for two of the most common shapes that can be used to build the area moment of inertia for many other shapes.

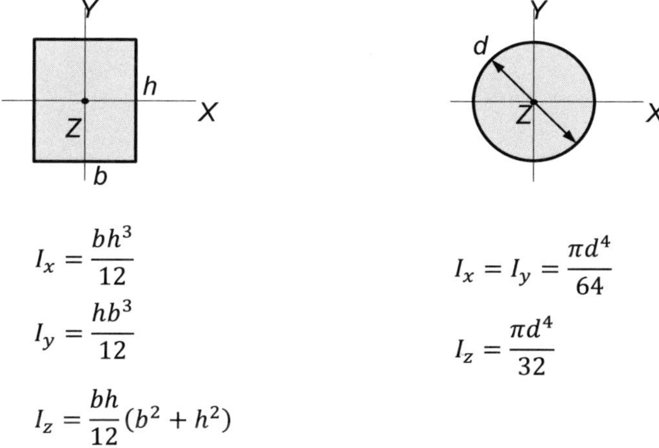

$$I_x = \frac{bh^3}{12}$$

$$I_y = \frac{hb^3}{12}$$

$$I_z = \frac{bh}{12}(b^2 + h^2)$$

$$I_x = I_y = \frac{\pi d^4}{64}$$

$$I_z = \frac{\pi d^4}{32}$$

$I_z = J =$ Polar Moment of Intertia

Figure 3.13 Area moment of inertia for common shapes.

If the beam shown in Figure 3.9 also has a tension load applied to it as shown in Figure 3.14, the direct tension stress can be determined from

$$\sigma_d = \frac{P}{A} \tag{3.143}$$

where σ_d = direct stress, P = direct load, A = cross-sectional area, and $\sigma_d = \frac{T}{b \times h}$.

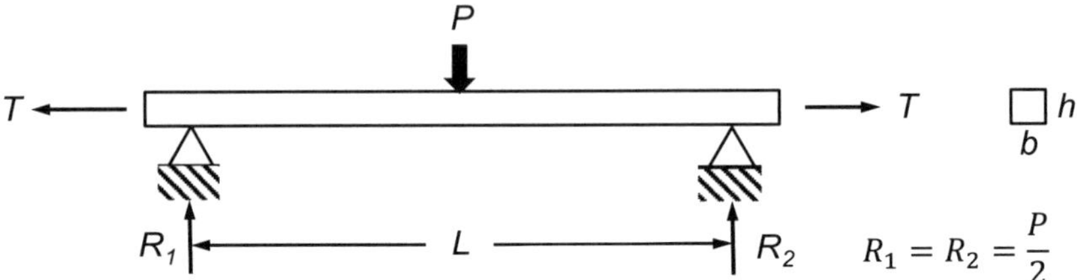

Figure 3.14 Beam with bending moment and tension load.

To get the true picture of stress state in a structural member, all the stresses need to be combined. There are various methods to combine the different stress states with the most common being the maximum shear stress theory, maximum/minimum principle stress, and Mises–Hencky stress theory.

The relationship between shear stress and bending and/or direct stress in the x and y directions and shear stress is shown by Mohr's circle. For the conditions stated, the stress state near the center of the beam in Figure 3.14 can be represented by Mohr's circle shown in Figure 3.15.

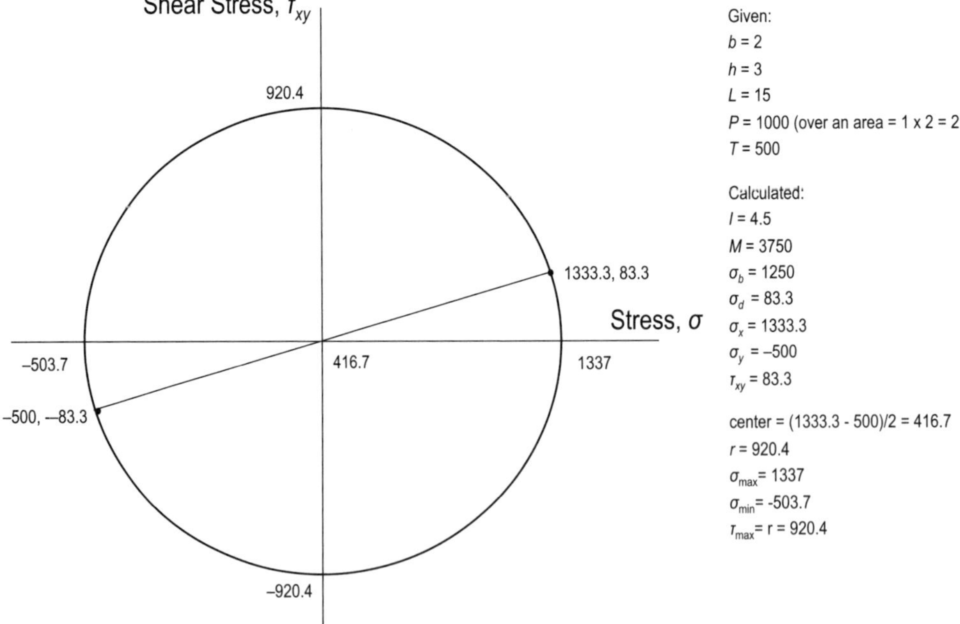

Figure 3.15 Mohr's circle for beam under combined loading.

The maximum shear theory of failure is used for ductile materials because they fail by shear that occurs at 45° to the principle stresses.

Therefore

$$\tau_{\max} = \frac{1}{2}\sigma.$$

For a combined stress state, maximum shear theory of failure is given by

$$\tau_{\max} = \frac{\sigma_1 - \sigma_2}{2} = \sqrt{\left(\frac{\sigma_x - \sigma_y}{2}\right) + \tau_{xy}^2} \quad (3.144)$$

For the conditions presented in Figure 3.15, the maximum shear theory of failure gives

$$\tau_{\max} = 920.4.$$

The Mises–Hencky theory of failure or distortion energy theory states that the equivalent stress state follows:

$$\sigma_e = \sqrt{\sigma_x^2 + \sigma_y\sigma_x + \sigma_y^2 + 3\tau_{xy}^2} \quad (3.145)$$

For the conditions presented in Figure 3.15, the Mises-Hencky theory of failure gives

$$\sigma_e = 1176.$$

A variation of this for a triaxial stress state is the von Mises or distortion energy theory, which is shown in (3.146). This form is often used to combine principal stress values in finite element programs:

$$\sigma_e = \sqrt{\frac{(\sigma_1 - \sigma_2)^2 + (\sigma_2 - \sigma_3)^2 + (\sigma_3 - .\sigma_1)^2}{2}} \quad (3.146)$$

where σ_1 = principal stress 1, σ_2 = principal stress 2, and σ_3 = principal stress 3.

The three principal stresses act in the three mutually perpendicular planes of zero shear. [Spotts 1978, 43–47, 101–106], [Salmon and Johnson 1980, 45–47]

3.4.2 Ultimate Strength and Plastic Analysis

While most designs utilize yield strength as the failure criteria, a structural component actually fails in tension or compression when the ultimate strength of the material is exceeded. However, this can be somewhat misleading because a ductile material will start "necking down" as the load increases from yield, so while the ultimate tensile stress may appear to be significantly larger that the yield strength, the actual load that can be carried by the structure may not be higher by the same ratio $\left(\frac{\text{Ultimate Tensile Strength}}{\text{Yield Strength}}\right)$. Once the yield strength of the material is exceeded, the stress/strain curve is usually not linear, so linear analysis techniques are no longer valid. This can complicate analysis but opens the door for plastic analysis techniques. In plastic analysis, one assumes that the entire stress field is at yield so the triangular stress field shown in Figure 3.10 becomes rectangular, as shown in Figure 3.16.

Here we introduce the concept of section modulus as the area moment of inertia divided by the distance from the neutral axis to the outer fiber as shown in

$$S = \frac{I}{c} \tag{3.147}$$

where S = section modulus, I = area moment of inertia, and c = distance from the neutral axis to the outer fiber.

For the rectangular beam of Figure 3.11, this becomes

$$\begin{aligned} S &= \frac{bh^3/12}{h/2} \\ S &= \frac{bh^2}{6} \end{aligned} \tag{3.148}$$

The plastic modulus is defined as

$$Z = \int y\,dA \tag{3.149}$$

For the rectangular beam of Figure 3.11, this becomes

$$\begin{aligned} Z &= 2\int_0^{h/2} yb\,dy \\ Z &= \frac{bh^2}{4} \end{aligned} \tag{3.150}$$

Comparing the section modulus S with the plastic modulus Z, we see that plastic analysis provides 50% more moment capacity:

$$\frac{Z}{S} = \frac{bh^2/4}{bh^2/6} = \frac{6}{4} = 1.5\,.$$

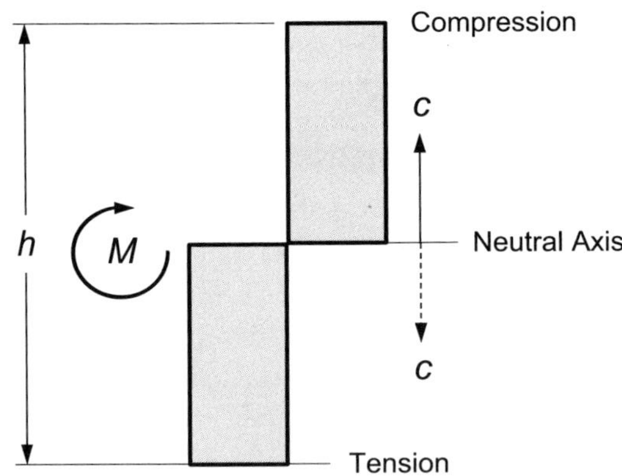

Figure 3.16 Bending moment stress field (plastic).

For the design and conditions shown in Figure 3.15, the bending moment is 3750 for an elastic bending stress of 1250. If the plastic condition of Figure 3.16 is reached at a stress of 1250, the plastic design moment is 1875. For the simply supported beam shown in Figure 3.9, the allowable load that can be applied according to plastic design theory is also larger by a factor of 1.5 or 1500. [Salmon and Johnson 1980, 339–341]

3.4.3 Fatigue Strength

For a design subjected to cyclic loading, the expected life and load profile should be defined as part of the requirement set. If the number of cycles at high load is low, then one can use the maximum load and design up to the yield strength of the material. If the loads are steady and do not fluctuate much so that the number of load cycles is low, one may also consider plastic analysis techniques. However, if the design is expected to survive a high number of fluctuating loads, then the fatigue strength of the material needs to be considered. The easiest approach is to design the system so that everything has a maximum stress below the material endurance limit. Unfortunately, some materials, such as aluminum, do not have a well-defined endurance limit. Also, the endurance limit is influenced by manufacturing processes. For example, the endurance limit for ground and polished wrought steels may be around 50% of the ultimate strength, while it may be only 35% to 40% of the ultimate strength for wrought steels with a machined surface, and even lower for materials in a hot rolled or forged condition. When a design under fatigue loading cannot be designed to the material's endurance limit and satisfy other design requirements such as cost, weight, and space claim, then a more refined approach is needed. [Spotts 1978, 97–100]

To perform a more detailed fatigue analysis, one needs to know the minimum and maximum loads that can be used to calculate the stress ratio, R. The stress ratio can range from -1 for a fully reversing load ($\sigma_{min} = -\sigma_{max}$) to 1 for a static load ($\sigma_{min} = \sigma_{max}$). One also needs to calculate the mean load and determine the load range along with the accompanying number of cycles for that fluctuating load. One should remember that the minimum and maximum loads may not represent the lower and upper bounds for the fluctuating load with the greatest number of cycles, respectively. Information from this more detailed analysis can then be combined with more detailed material information from a stress-life or S-N curve so that stress ratio is given by:

$$R = \frac{\sigma_{min}}{\sigma_{max}}. \tag{3.151}$$

A stress-based approach to understanding the fatigue life of a cyclically loaded design can be gained from a constant life fatigue diagram. A constant life fatigue diagram is specific to a material and material condition. This approach assumes deformations are elastic, that is, local plastic strains are neglected. A generic constant life diagram is shown in Figure 3.17.

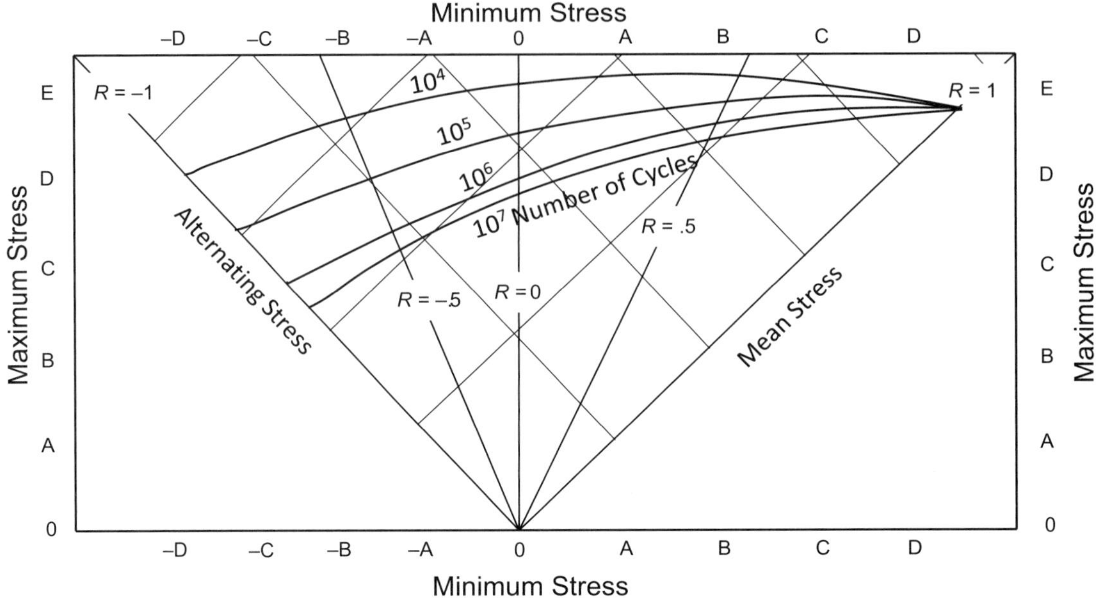

Figure 3.17 Generic constant life diagram.

In a strain-based approach to understanding fatigue life of a cyclically loaded design, both elastic and plastic deformations can be characterized. This diagram is similar to an S-N curve with a log-log plot of the number of strain reversals (two times the number of cycles) in the *x*-axis and the strain amplitude (half the strain range) in the *y*-axis (Figure 3.18).

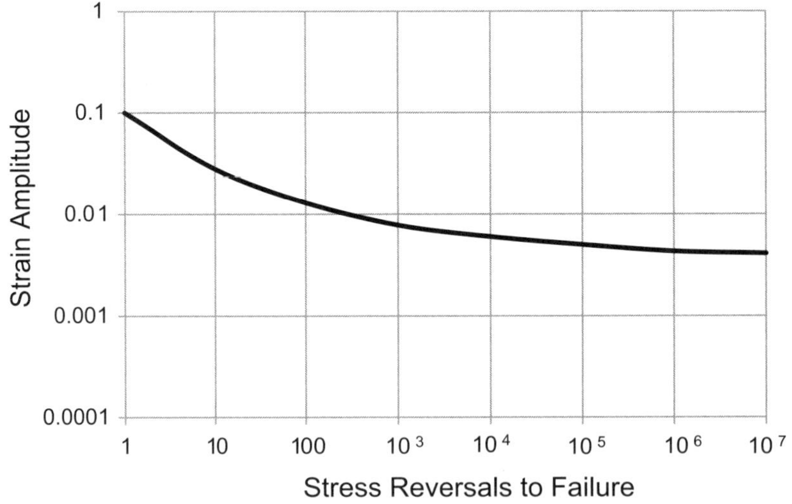

Figure 3.18 Generic strain versus strain reversal diagram.

The endurance limit is defined as the maximum stress that a material can sustain for infinite life. If the stress is higher than the endurance limit, then the design is not expected to have an infinite life. A plot of stress and life in the number of cycles is referred to as an S-N curve. S-N curves are material dependent and plot the number of cycles to failure in a log scale in the *x*-axis and the stress in the *y*-axis (Figure 3.19). This information can then be used to refine the design so stresses to align with the required life.

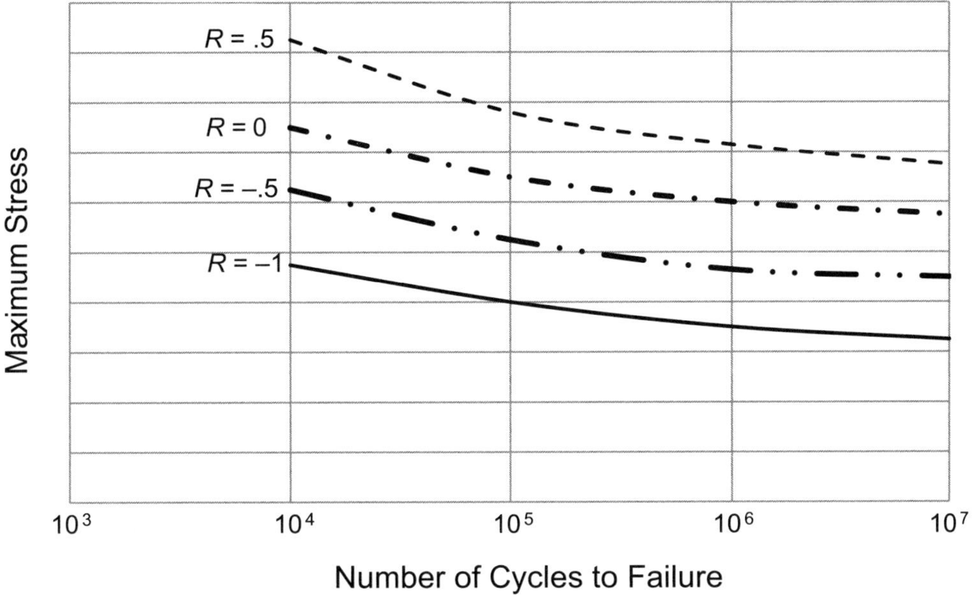

Figure 3.19 Generic S-N curve.

Another way to look at the relationship between the alternating stress due to cyclic loading and material strength is with the Goodman diagram (Figure 3.20). This is a somewhat simplistic representation of the relationship but can provide insight into what is happening to a part due to the load profile. The Goodman relationship is shown in the following and with the resulting diagram in Figure 3.20:

$$\frac{\sigma_a}{\sigma_e} = 1 - \frac{\sigma_m}{\sigma_{UTS}} \tag{3.152}$$

where σ_a = alternating stress, σ_e = endurance limit, σ_m = mean stress, and σ_{UTS} = ultimate tensile strength. [Harris and Crede 1976, 42-34 to 42-35]

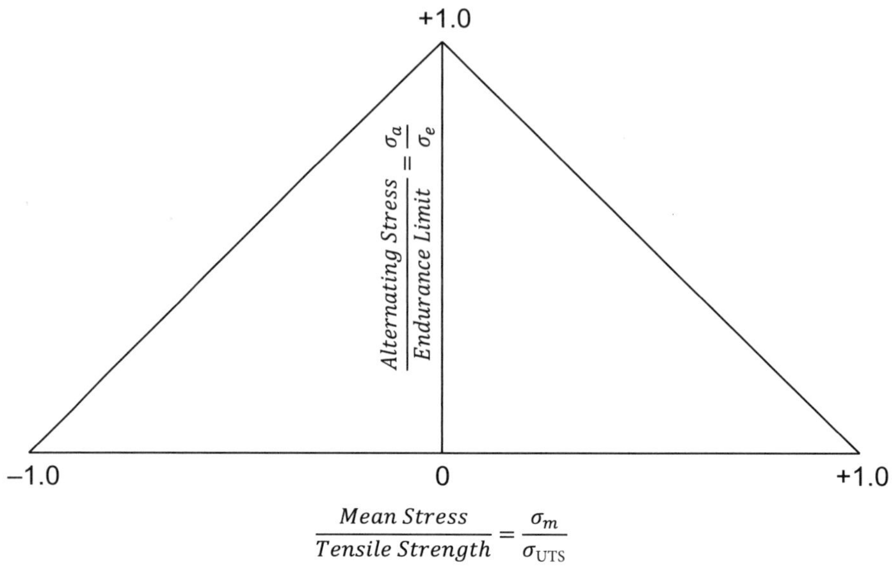

Figure 3.20 Goodman diagram.

When there are more than one stress level in a design's life that is above the endurance limit, then the engineer needs to combine the all the incremental load effects on life to determine the total number of cycles that the design can survive without failure. Miner's equation is typically used to determine the life of a design subjected to combined stresses over the endurance limit:

$$\frac{L_1}{N_1}+\frac{L_2}{N_2}+\frac{L_3}{N_3}+\cdots+\frac{L_x}{N_x}=\frac{1}{N_c} \tag{3.153}$$

where L_x = portion of the time spent in fully reversed cycles at a stress level of σ_x, N_x = life in number of cycles for a stress of σ_x, and N_c = combined life of the design in cycles. [Spotts 1978, 11–113]

For example, if a design spends 20% of its life at a stress level that has a life of 10,000 cycles, 30% of its life at a stress level that has a life of 40,000 cycles, and 50% of its life at a stress level that has a life of 60,000 cycles, its combined expected life would be cycles

$$\frac{.20}{10{,}000}+\frac{.30}{40{,}000}+\frac{.50}{60{,}000}=\frac{1}{N_c}$$

where N_c = 27,907.

Other material characteristics to consider when determining the strength of a design include material elongation, toughness or impact strength, hardness, and corrosion resistance. Many engineers like to use materials with a minimum elongation of 10% to ensure that the material behaves elastically and any discontinuities can be absorbed by the design. Impact strength (toughness) considerations are similar, to ensure the component or components have the required combination of strength and ductility to manage the impact energy that can come from operating, mobility, transportation, and/or other impact events. Toughness of a material is often defined from Charpy or Izod impact tests, defined in ASTM E23. In these tests, the energy required to break notched specimens with a pendulum is measured. This impact energy is usually obtained to specific materials through a material testing facility. [Boyer and Gall 1985, 34.18–34.21] Material hardness is important to the designer due to its relationship with material strength. For example, harder steels have a higher tensile strength than steels with lower hardness. Hardness also relates to the ability of a material to resist wear, which could cause it to fail through material loss. Higher material hardness can also help a material resist nicks and dents that can produce stress concentrations, which can become failure initiation sites. The ability of a material to resist corrosion can also determine how well it maintains strength through its intended life. The effects of corrosion can be mitigated by protective coatings; however, the impact of the coating on function and cost needs to be evaluated. As in all design activities, the designer needs to be aware of the limitations of analysis conducted, material data, and load information to apply an adequate safety margin to the design.

3.5 Constraints

Just as the initial design needs to consider the desired payload, operating timeline, and potential material properties, it also needs to consider the constraints imposed on the actuator, structure, and other associated components. These constraints may have more influence on the final design than the operating parameters. That is why they play an important role in defining the initial design generated during requirements analysis. If they are not considered, the initial concept may be deficient and require unplanned effort to implement all the modifications needed to satisfy the constraints. In the worst case, the initial design concept needs to be discarded and a new concept generated that can satisfy all the constraints. This may cause the project to miss budget and schedule goals, jeopardizing the entire program. Constraints may include mounting configuration, environment, safety, reliability, maintenance requirements, cost, and risk. While the latter two, cost and risk, are influenced by the design, they are part of the business case for any project and therefore are covered as part of program management.

3.5.1 Mounting Configuration

Actuator mounting needs to be understood so that the actuator can be designed and analyzed accurately. In most cases, actuators do not like to inherit moments from the load or structure, so utilizing mounting techniques that limit the possibility of transferring a moment to the actuator is a good idea. In analysis terms, this means pinned joints and simply supported members are good, but fixed supports are not. As for the load path and how it makes its way through the structural members is assessed, the designer should make sure all loads are considered. Direct actuator loads certainly need to be considered, but one should also make sure the design accounts for shock and vibration loads. Shock and vibration loads may come from exciting structural modes during operation or externally from transportation, mobility, or operation of nearby equipment.

3.5.1.1 Pinned Joints

Pinned joints limit the possibility of transferring moments in the pinned planes but can transfer them through the remaining plane unless a double pinned joint is used. In some cases, allowing the joint to rotate all planes is easy to incorporate into the design, but the engineer must be aware of the need. If a linear actuator uses pins through spherical bearings at both ends, only tension or compression loads can be transferred through the actuator. This can provide one of the best loading conditions for a linear actuator because it needs only to accommodate loads in the line of action, which it is primary design condition. Like linear actuators, there is a desire to limit bending moments in the shaft of a rotary actuator to keep bearing loads manageable and minimize the space required to counteract the moment loads. One easy way to limit bending moments in shafts is to use universal joints, constant velocity joints, or some type of coupling. Use of these components can also accommodate misalignment that may exist between the rotary actuator and the rest of the drive system. As the design progresses, it may be necessary to account for friction in the pinned joints to make sure the actuator and connecting hardware can handle any friction induced moments.

3.5.1.2 Simply Supported Joints

In most cases, simply supported joints do not lend themselves to actuator design because more control over the behavior of the connection is needed to ensure safe, stable operation. Simply supported joints are one of the first supports that the beginning engineer learns to analyze. They are easy to analyze because they do not transfer moments or side loads to the actuator, but have the drawback that they can only provide a compressive reaction load.

3.5.1.3 Fixed Support Joints

As by definition a fixed joint is "fixed" and does not move, it can only be utilized on one side of an actuator—the side that does not move. Fixed supports can transfer linear and moment loads from all directions to the actuator. This can provide a very stiff joint and allows complete control of the loads but also requires the actuator to handle all these load conditions. As the joint does not automatically limit the load going to the actuator, the engineer must determine and evaluate all the possible loads. In some cases, it is hard to identify all the applicable loads and then determine their magnitude. In spite of the analytical challenges with fixes supports, sometimes they are the best type of joint for the design and therefore the disadvantages outweigh the burdens.

3.5.1.4 Joint Combinations

To satisfy concept requirements, sometimes a combination of joint types is selected for the actuator. Any combination can work, but the engineer needs to return to engineering fundamentals and make sure that all the loads are balanced. A good tool for this activity is a free-body diagram of the actuator and its connecting hardware. One of the most common combinations is pinned on one end and fixed on the other, which allows the actuator to be rigidly fixed to the supporting structure and then connected to the load mechanism through a pinned joint. This can be a successful combination if all loads and moments are carefully analyzed and then balanced in the design. Sometimes this means extra bearings in the actuator, additional support structure, or extra material in the actuator components. For custom actuators, these features may not be an issue, but they may be hard to get incorporated into a standard commercial component. The specific details of the design do not need to be completed in the initial design concept, but the general approach of using custom versus standard commercial components needs to be addressed.

3.5.2 Environment

Environmental considerations have two components: the effect of the environment on the product and the product's effect on its environment. These environmental interactions need to be understood and features incorporated into the design to allow the product to successfully survive, operate, and not harm the environment it has to interact with. Environmental effects on a product are often split into two categories: a less severe operating environment and a more severe storage or nonoperating environment. Several specifications such as MIL-STD-810 (environmental engineering considerations and laboratory tests) and RTCA/DO-160 (environmental conditions and test procedures for airborne equipment) contain valuable environmental condition and approved test methods to guide the design of products. A product's negative effect on its environment is typically pollution of some sort, which occurs in the manufacturing process, during operation, or at the end of life disposal. As with all potential design constraints, the engineer should make sure that environmental constraints are captured in the requirement set as a way to communicate them to the customer and other engineers.

3.5.2.1 Operation

Some products operate inside an enclosure where the environment is controlled. In this scenario, environmental factors do not have a large impact the design. Other products are expected to operate with little degradation over broad temperature ranges, precipitation conditions, and other adverse conditions. For example, some of the outdoor machineries need to operate through a temperature range of −40°C to +49°C or greater. There is also rain, mud, snow, ice buildup, dust, humidity, and solar loading that should be considered during the design of components exposed to the outside environment. However, the engineer should be aware that the interior environment can be worse than the exterior environment if the operating space is not conditioned.

3.5.2.2 Storage

Depending on the actuator's operating environment, its storage environment can be broaderor narrower than its operating environment. For ground equipment, storage cold temperatures may be lower than their operating range, but the hot temperatures may not. For aerospace applications, the expected operating range can be very broad, so storage temperature considerations do not drive the design. If storage temperatures have a broader range than the operating temperature, the design approach and material selection may be adjusted because most performance parameters do not need to be satisfied when a product is stored. However, if this is the case, the actuator must survive storage and return to a fully operating state with no degradation when removed from storage.

3.5.2.3 Shipping and Transportation

Shipping or product transportation loads need to be considered during the design process. These loads can induce fatigue damage in a system before it is even placed in service. While the damage may not be apparent when the system is placed in service, it can fail prematurely due to the damage it received during shipping or transportation. Complete characterization of transportation requirements can drive this part of the design process to make sure that the loads are considered. Shipping or transportation loads are often vibration and shock loads that may also excite structural modes of the system. If these loads add significant cost and/or complexity to a system, they may want to be managed with carefully designed packaging or shipping containers. Additional load cases may be identified as part of the design, analysis, and structured review process such as Design Failure Mode and Effect Analysis (DFMEA). It is part of the product design engineer's job to ensure that the product is not damaged by the time it gets to the final customer.

3.5.3 Pollution

The engineer needs to anticipate how the product is going to interact with its environment so that pollution is minimized. Ideally, this interaction encompasses the product's entire life cycle from manufacturing the first piece through product decommissioning. Through education and training, the engineer has the knowledge and ability to obtain information on manufacture, assembly, test, product use, maintenance, failure modes, and decommissioning. This information needs to be collected and documented, so informed decisions can be made to minimize the product's environmental impact while meeting all other design requirements. The designer needs to work with a well-established material expert or perform the necessary research to understand the acceptance of the selected materials in terms of current and future regulations. The success of the design does not want to be dependent on a material that is going to become unavailable or forbidden.

3.5.4 Safety

During the requirements analysis stage, specific safety considerations need to be identified and considered as part of developing the initial design concept. In general, there are three areas that need consideration: personnel safety, machine safety, and environmental safety. The first needs to address all potential hazards against operators and bystanders. The second needs to address all hazards that may cause the machine to damage itself and some other machine or structure. The third addresses potential hazards that the machine may pose to the environment. Requirements need to communicate the importance of each of these to the engineer so that they can be appropriately incorporated into the design. Personnel safety is usually the most important followed by machine safety with degradation occurring in a manner that minimizes pollution. In all cases, the appropriate design margin must be utilized so that unknown factors do not cause a hazardous condition. Identifying hazards and safety issues can come out of design failure mode and effects analysis (DFMEA), failure mode and effects analysis (FMEA), and failure mode, effects, and criticality analysis (FMECA).

3.5.4.1 Personnel Safety

Ensuring the safety of all personnel around and being involved with the equipment is one of the design engineer's primary responsibilities. As the design engineers developed, analyzed, and designed a piece of equipment, they should be able to identify potential failures and the effect of failures better than anyone else. If the system does not fail to a safe condition, the design should be changed so that hazards to personnel are minimized.

3.5.4.2 Machine Safety

It is important for design requirements to reflect the intent for machine safety. If the machine cost is low enough so that it is considered to be disposable, there may not be a need or desire to make sure that it does not destroy itself during a failure event. However, if personnel are depending on the machine for their safety, then it is important that the machine fails gracefully. In this case, the machine impacts personnel safety indirectly, so maintaining limited operation or warning of imminent failure is important to maintain personnel safety. In other cases, machine cost is high, so adding features that limit the level of damage it does to itself can be cost effective. Finally, customer perceptions on machine failures are an important consideration. A machine that fails gracefully makes better impression on current and future customers than the one that fails to a mangled mess.

3.5.4.3 Environmental Safety

Concerns on how equipment failure affects the environment need to be clearly defined in the requirements. Minimizing environmental impact is always good practice because it minimizes cleanup costs and external environmental protection features, and is just being environmentally responsible. In some cases, failure of an actuator can have a direct impact on its immediate environment such as when a hydraulic cylinder fails, leaking or spraying hydraulic fluid all over. In other cases, the effect can be indirect such as when the same hydraulic cylinder fails, but instead of leaking hydraulic fluid, it causes a tank of hydraulic fluid to rupture, again leaking hydraulic fluid, but at a location remote from the cylinder. In either case, a more environmental friendly failure would be automatically detected and flow to the cylinder be stopped. Environmental safety could also include high noise levels if the machine operates or fails in such a way that it creates a noise hazard. This is not to say that failures cannot or should not result in additional noise that offers the benefit of indicating an abnormal condition to operators and/or other personnel.

3.5.4.4 Safety Summary

Actuators provide the action for machine motion, so how they fail or act when another component fails has a significant impact on how the entire machine fails. Therefore, reviewing actuator failure modes and actuator actions during the failure of other components is an important part of developing a safe product. During the requirements analysis phase, the initial concept needs to reflect the safety goals of the customer. An FMEA, FMECA, DFMEA, or DFMEA may not be performed at this stage, so it is the responsibility of the engineer to make sure personnel and equipment safety goals stated in the requirements can be implemented in the design. Also, if the requirements do not contain adequate safety requirements, the engineer should make sure that additional safety requirements are added. This allows the customer to understand any potential increase in design scope, documents the safety expectations for other engineers and managers, and makes sure that the initial design concept incorporates all the important safety features.

3.5.5 Reliability

It is obvious to most people that reliability is important to users and customers. However, higher costs generally accompany an increase in reliability. High reliability also often conflicts with high performance and maintenance ease. Therefore, the challenge for engineers is to create a design that achieves reliability goals while maintaining performance, cost, and maintenance goals. If reliability is not considered during the initial design phase and reflected in the final requirement set, the opportunity to make a meaningful impact on reliability may be lost. Often, reliability improvements made in later stages of the design process respond to reliability issues and add cost and complexity with limited reliability improvement. As actuators represent some of the more complex elements in a machine and are subject to wear, they represent or affect a significant portion of a machine's overall reliability. While significant performance improvements may require new technologies, reliability and cost are generally enhanced by utilizing more mature technologies. Therefore, a good approach is to leverage knowledge from mature technologies and apply them to new technologies being used to improve performance. The reliability of a system or machine can be captured in the mean time between failures (MTBF) and mean cycles between failures (MCBF), or failure rate. The numbers from these calculations provide a metric to compare the reliability of various parts, subsystems, or systems, but as with all such numbers, the calculation needs to be understood so that it does not produce a misleading result. Sometimes the customer is interested in the reliability relative to a type of failure such as critical failures or failures that cause system aborts:

$$\mathrm{MTBF} = \frac{\text{Total Time in Measurement Period}}{\text{Number of Failures}} \tag{3.154}$$

$$\mathrm{MCBF} = \frac{\text{Total Cycles in Measurement Period}}{\text{Number of Failures}} \tag{3.155}$$

where the number of failures can be any of the following:

- all failures when determining overall repair needs
- all critical failures when determining system availability

- all failures that cause system aborts when assessing the likelihood that a task or mission can be completed

$$\text{Failure Rate} = (\text{BFR}) \times C_1 \times C_2 \times \ldots C_n \tag{3.156}$$

where BFR = base failure rate for the design (failures/unit of measure) and $C_1 \times C_2 \times \ldots C_n$ = product of all factors affecting the BFR in the application.

During the requirements analysis and concept formulation stage of the design cycle, engineers are usually asked to assess the design to make sure that it is on a path to satisfy stated reliability requirements. This is usually done by comparing the design with some known legacy system and then factoring in design maturation from similar development projects. If no existing products are available for comparison, the designer needs to look at individual components to see how their combined reliability relates to the goals. All of this requires time and effort, which is usually in short supply so that all assumptions made during the process need to be documented so they can be updated as the design matures.

3.5.6 Maintenance

Maintenance can fall into two categories: preventative and repair. Customers tend to dislike both because they remove the machine from a productive state, require labor and materials, and can be a source of additional problems. However, of the two, repair is more of an unplanned event, so it is disliked the most. It can leave people stranded, without product and even in danger, all of which create an unpleasant memory for the user and customer. This is not what the engineer or manager wants, so most organizations opt for some sort of preventative maintenance program. It is important to understand the maintenance philosophy and have this documented during the requirements analysis phase. This keeps designers on track to achieve the maintenance objectives throughout the design phase, so issues do not suddenly appear during prototype or production phases.

3.5.6.1 Preventative Maintenance

While developing the initial design, engineers need to incorporate features and ideas that are consistent with the preventative maintenance philosophy as defined in the requirements. If the requirements do not address preventative maintenance, it is the engineer's responsibility to define a preventative maintenance philosophy and document it as part of the requirement set. The preventative maintenance philosophy may be structured around performing maintenance operations that protect or improve product life, much like changing the oil in one's car. It can incorporate changing parts, similar to spark plugs, before they fail if they have a known or predicted life in similar applications. Another approach may be to replace parts as they are about to fail as determined by monitoring equipment/sensors or visual inspection. Examples of this latter condition may be visual inspection of worn tires, initiation of fatigue cracks in structural members, noise from wear sensors on brake pads, or vibration sensors that indicate imminent bearing failure. The final preventative maintenance philosophy may be a combination of these approaches that satisfies the life, safety, cost, and operational needs of the equipment.

Addressing maintenance early in the design cycle may seem like unnecessary effort, but failure to address it then only pushes the problem further down the development cycle. If this happens, the designer and customer's expectations can continue to diverge to a point where significant design changes are required to get them realigned. Some examples of design decisions that impact maintenance include using items such as self-lubricating bearings, sealed bearings, fluid selection, filterselection, sensors to detect component health, and component replacement versus component repair. In all cases, it is important for the engineer to understand what the preventative maintenance items are and then configure the design so that maintenance can be performed easily. Eliminating maintenance actions through design also eliminates the potential for maintenance-induced failures. Where maintenance cannot be eliminated, it should be made easy and unobtrusive. The more disassembly involved, the greater the chance of breaking, losing or damaging some other component. Also, more effort is needed to maintain a system that requires significant disassembly and reassembly during maintenance. A maintenance ratio (MR) may be computed to express how much effort a system requires to keep it operating at peak performance levels. Generally, the MR is defined as the number of maintenance hours required divided by the operating hours, so a low number is good. Often the engineer is asked to estimate the MR to determine if the proposed design is on track to satisfy customer expectations. This can be hard to accomplish and requires assumptions during the design concept phase, but often the design engineer is the only person that can do it. The engineer needs to remember to document all assumptions made so that they can be verified or updated as the design matures.

3.5.6.2 Repair

Repair is usually in response to some sort of failure. In the mechanical world, failures are generally the result of a part breaking or wearing out. If wear is well understood by the engineer, it can be incorporated into a preventative maintenance plan. This allows customers to replace the part at their convenience rather than when wear causes the entire machine to stop working. This is often at the worst time and can also cause other components to fail. Part of the initial design should include a disassembly plan along with an idea of how to replace and/or repair various components. The ease with which this can be accomplished is sometimes captured in the mean time to repair (MTTR), which is the average time it takes to repair the machine. While this number gives an indication of how easy an item is to repair, its calculation needs to be understood because one way to have a low MTTR is to require a lot of repairs. This is of course undesirable because it indicates the system also has a low MTBF.

While people like to discuss, assess, and compare the various MTBF and MTTR numbers, often the overall availability of a machine is what really matters. Availability is a combination of MTBF and MTTR and gives the customer or user an indication of the real cost of ownership from a reliability and maintenance viewpoint. If is also useful to understand if a machine is going to be "available" to do a task when called upon.

3.6 Verification and Validation

This is a good place to make sure the reader is clear on the meaning of verification and validation. Relative to design and design requirements, verification is used to confirm that the design is stable, backed up with relevant analysis and test data, and satisfies the design requirements. If requirements are well written, with specific pass/fail criteria, verification steps are clear and objective. Validation assesses how well the design satisfies the customer's needs. Therefore, validation is a more subjective process, with the customer, end users, and others determining if the design meets the customer's needs. If the requirement set is done well, passing all the verification steps means that final product validation should go smoothly.

In the requirements analysis phase of a project, verification and validation revolve around the analysis techniques and tools to reduce project risk, building confidence in the actuator's ability to satisfy customer requirements. Validation of the analytical models and tools used to predict the mechanical behavior of the actuator system (as determined by performance, strength, and stiffness analyses) is the key to developing confidence in the design they represent. Tool and model validation can be based on sound fundamentals that have been proved over time and correlated with test results. Depending on the uniqueness of the design, test results can come from preprototype concept testing and subscale testing or be parented from tests performed on similar designs. Either way, analysis techniques with proven correlation to test results offer all members of the design and management team a way to accurately and confidently predict a high-performance actuator's behavior and reduce project risk. These validated tools can then be used to provide analytical evidence to verify the design against the defined requirement set.

3.7 Summary

Requirements analysis is more than just listing a set of requirements. It needs to represent the customer's needs and a valid design concept. For actuators, the requirements analysis effort should evaluate performance, loads, strength and stiffness, and design constraints. The best solution is usually the simplest design that achieves these goals. Complexity to address performance, safety concerns, or enhanced maintenance can add cost and weight, while decreasing reliability. Therefore, the design concept selected for refinement during subsequent design phases should reflect a design that carefully balances all requirements.

3.8 References

Byars, E. F., and Snyder, R. D. 1975. Engineering Mechanics of Deformable Bodies, Third Edition. Intext Educational Publishers: New York.

Shortley, G., and Williams, D. 1971. Elements of Physics, Fifth Edition. Prentice-Hall, Inc: Englewood Cliffs, NJ.

Swokowski, E. W. 1975. Calculus With Analytic Geometry. Prindle, Weber & Schmidt, Inc: Boston, MA.

Spotts, M. F. 1978. Design of Machine Elements, Fifth Edition. Prentice-Hall, Inc: Englewood Cliffs, NJ.

Naval Air Systems Command, Naval Sea Systems Command, Naval Supply Systems Command, Space and Naval Warfare Systems Command, Marine Corps Systems Commnad. 2004. Naval Systems Engineering Guide. Department of the Navy: Washington, DC.

Reswick, J. B., and Taft, C. K. 1967. Introduction to Dynamic Systems. Prentice-Hall, Inc: Englewood Cliffs, NJ.

Harris, C. M., and Crede, C. E. 1976. Shock and Vibration Handbook. McGraw Hill Book Company: New York.

Meriam, J. L. 1975. Statics. John Wiley & Sons, Inc: New York.

Salmon, C. S., and Johnson, J. E. 1980. Steel Structures Design and Behavior. Harper and Row: New York.

Boyer, H. E., and Gall, T. L. 1985. Metals Handbook Desk Edition. American Society for Metals: Metals Park, OH.

Beer, F. P., and Johnston Jr., E. R. 1976. Mechanics for Engineers, Dynamics Third Edition. McGraw-Hill Book Company: New York.

Chapter 4
Design to Requirements

This phase is a continuation of the design process where the concept previously defined, backed up with written requirements, and accepted by the customer is refined and matured. This should be the fun part. The engineer can now concentrate on defining the details of the design, performing analysis to make sure it is going to perform as expected, and prepare for building the prototype. There are several steps in developing a design that meets the requirements and the process continues to be iterative, but developing a working solution can be very rewarding. Some of the steps in this phase may include defining the system's performance allocations, determining the power source, and finally determining the actuator type along with its size and other characteristics. This phase may be segregated into various steps, each of which concludes with a review. Successful passage of each review with the appropriate verification/validation evidence allows the project to proceed to the next step.

4.1 Performance Allocations

Depending on machine complexity and resolution of the requirements handed to the engineer, one of the first tasks may involve allocating the performance requirements to all actuators in the system. This is best accomplished by creating an operating cycle for the design concept and assigning times to each motion with proper overlaps or lags incorporated into it. Overlaps and running motions in parallel increases the risk of equipment collisions but also reduces cycle time without the need to shorten individual motion times. A notional cycle for moving an item from A to F is shown in Figure 4.1.

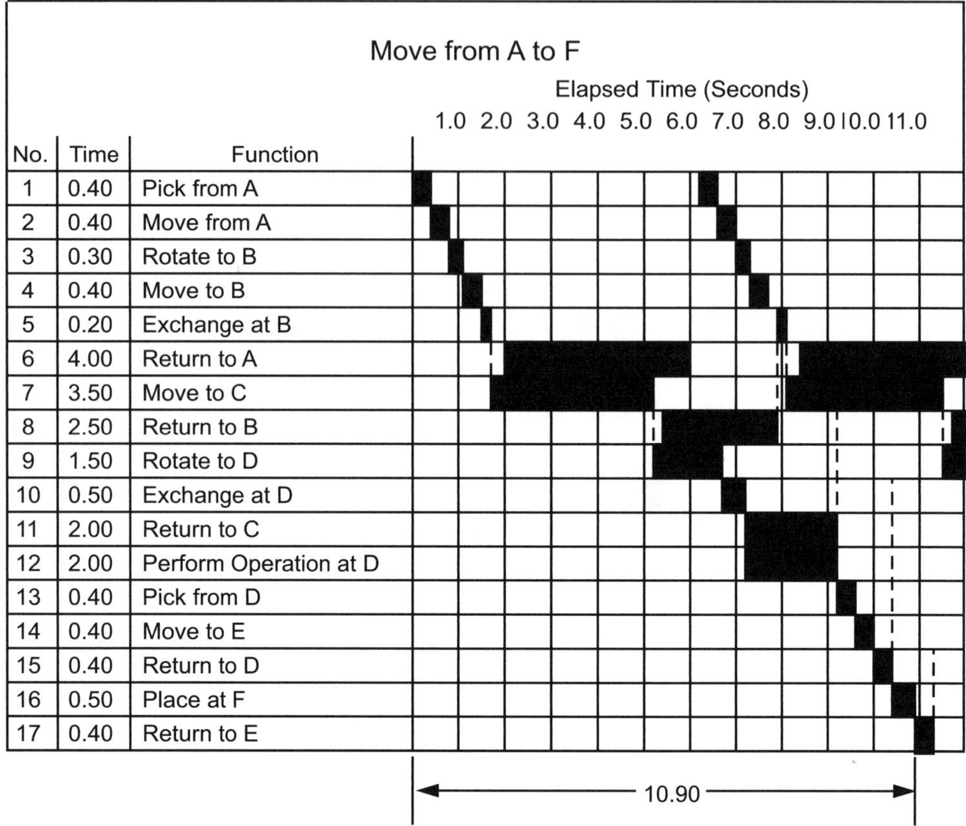

Figure 4.1 Simple time cycle.

The same amount of work may be done moving an object from A to F, but if accelerations can be reduced by overlapping and running parallel operations, then the overall energy consumed can be reduced and component sizes may be reduced. Figure 4.2 shows that the overall move from A to F can still be completed in 10.90 s while increasing the time for operations 1, 2, 4, 14, 15, and 16 if the overlap between operations 3 and 4, and 15 and 16 is implemented.

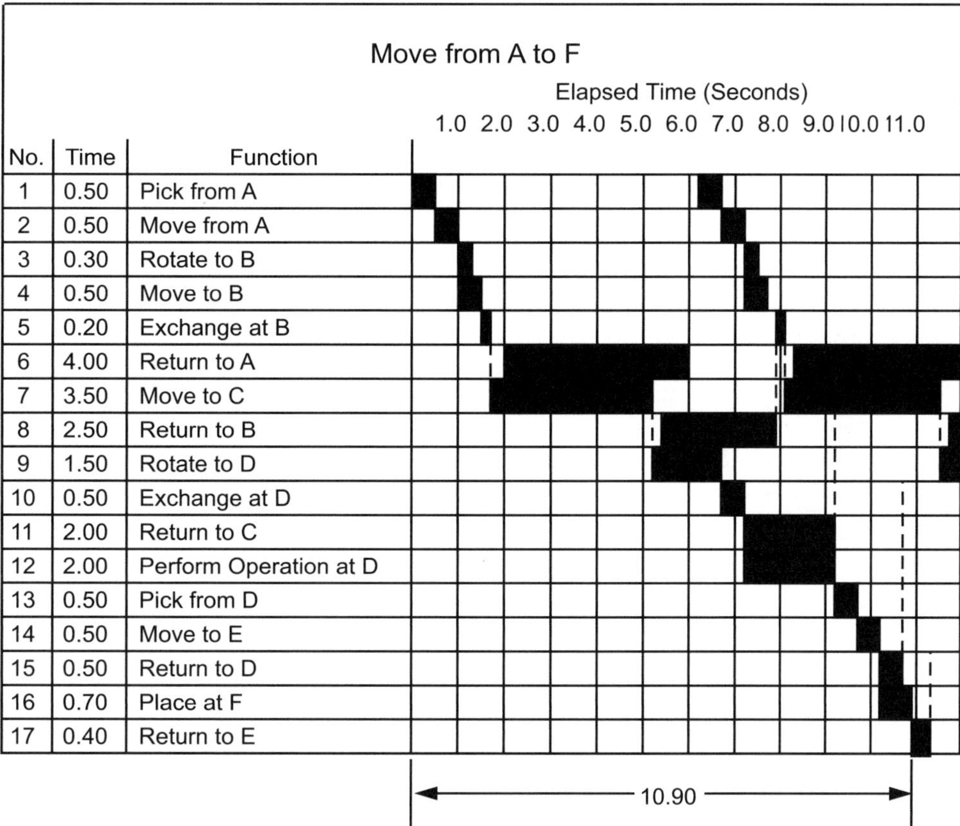

No.	Time	Function
1	0.50	Pick from A
2	0.50	Move from A
3	0.30	Rotate to B
4	0.50	Move to B
5	0.20	Exchange at B
6	4.00	Return to A
7	3.50	Move to C
8	2.50	Return to B
9	1.50	Rotate to D
10	0.50	Exchange at D
11	2.00	Return to C
12	2.00	Perform Operation at D
13	0.50	Pick from D
14	0.50	Move to E
15	0.50	Return to D
16	0.70	Place at F
17	0.40	Return to E

Figure 4.2 Time cycle with overlap.

If we utilize the principles of motion discussed earlier on operation number 2, we can see that acceleration may need to be increased substantially to achieve a 0.1-s reduction in move time. Figures 4.3 and 4.4 show the impact this change can make on the actuator size needed for the move.

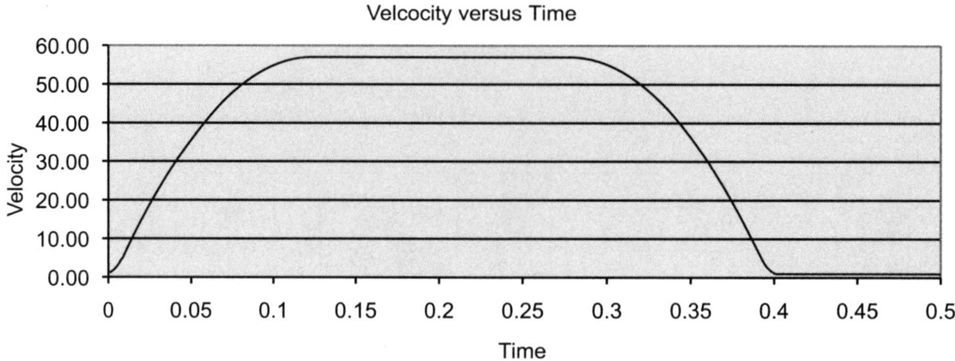

Figure 4.3 The velocity profile for simple time cycle, where distance = 18, time = 0.4, acceleration = 1000, deceleration = 850, maximum velocity = 57, natural frequency = 750, and jerk at maximum velocity transition = −8,000

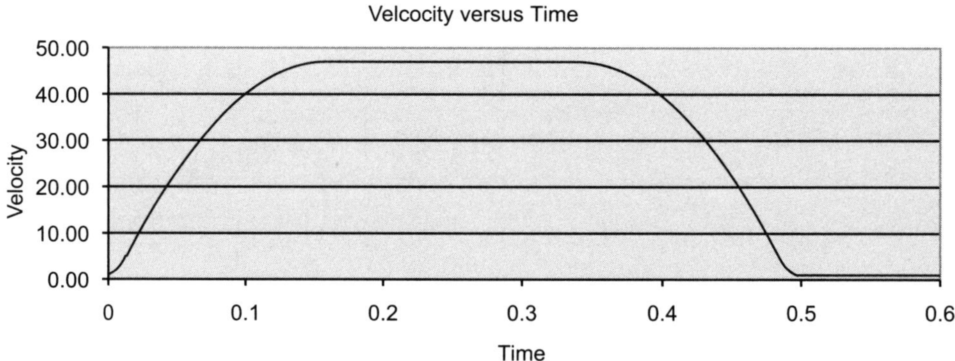

Figure 4.4 The velocity profile for time cycle with overlap, where distance = 18, time = 0.5, acceleration = 650, deceleration = 550, maximum velocity = 47, natural frequency = 500, and jerk at maximum velocity transitions = −4000.

From this example, one can see that the maximum velocity increases 20%, acceleration increases 55%, deceleration increases 55%, jerk increases 100%, and the natural frequency increases 50% just to decrease the move time .10 s. This 20% change in move time because overlap was not incorporated into the operating cycle has a major impact on the force (55% increase from acceleration), power (87% increase from force and velocity), and stiffness (125% increase from natural frequency) requirements. There are other combinations of acceleration and velocity that can reduce the move time by .10 s, but they all require an increase in power and stiffness:

$$\text{As } F = ma \text{ and } W = F \times x \tag{4.1}$$

$$W_1 = m \times a_1 \times x$$
$$W_2 = m \times a_2 \times x \tag{4.2}$$

$$\frac{W_1}{W_2} = \frac{m \times a_1 \times x}{m \times a_2 \times x} \tag{4.3}$$

$$\frac{W_1}{W_2} = \frac{a_1}{a_2} \tag{4.4}$$

$$P = F \times v \tag{4.5}$$

[Shortley and Williams 1971, 56, 102–121], [Beer and Johnson 1976, 496–498]

As $F = ma$ and $P = F \times v$

$$P_1 = m \times a_1 \times v_1$$
$$P_2 = m \times a_2 \times v_2 \tag{4.6}$$

$$\frac{P_1}{P_2} = \frac{m \times a_1 \times v_1}{m \times a_2 \times v_2} \tag{4.7}$$

$$\frac{P_1}{P_2} = \frac{a_1 \times v_1}{a_2 \times v_2} \tag{4.8}$$

$$\text{As } F = ma \text{ and } E_k = \frac{1}{2}mv^2 \tag{4.9}$$

$$\frac{E_1}{E_2} = \left[\frac{v_1}{v_2}\right]^2 \tag{4.10}$$

$$E_1 = \frac{1}{2}mv_1^2$$
$$E_2 = \frac{1}{2}mv_2^2 \tag{4.11}$$

$$\frac{E_1}{E_2} = \frac{\frac{1}{2}mv_1^2}{\frac{1}{2}mv_2^2} \tag{4.12}$$

$$\frac{E_1}{E_2} = \frac{v_1^2}{v_2^2} \text{ or } \frac{E_1}{E_2} = \left[\frac{v_1}{v_2}\right]^2 \tag{4.13}$$

Time spent refining the operating cycle and motion profiles can pay big dividends by creating a better understanding of the design, improving energy efficiency, reducing actuator size, and reducing structural requirements. Once performance allocations are in place for the system, the engineer can start performing other basic calculations to further define system components. Based on the overall move profile, the engineer should calculate how much work is being on the payload. This number along with the rate provides a baseline on how much power would be required in a perfect system. For example, if the move from A to F takes a 35-kg cylinder horizontally 3 m and up vertically 2 m every 10.9 s

$$\text{Work} = 0.1 \times 35\ \text{kg} \times 9.806\ \text{m/s}^2 \times 3\ \text{m} + 35\ \text{kg} \times 9.806\ \text{m/s}^2 \times 2\ \text{m}$$

$$\text{Work} = 789\ \text{N} \cdot \text{m}$$

$$\text{Power} = \frac{789\text{N} \cdot \text{m}}{10.9\text{s}} \times \frac{\text{W}}{1\ \text{N} \cdot \text{m/s}}$$

$$\text{Power} = 72.4\ \text{W}.$$

In reality, the path taken by the load requires acceleration and deceleration of the mechanism and the load for each move as so much more power and energy is required. Unless the energy can be captured during the deceleration phase, it is wasted and creates heat that ultimately has to be dissipated by the system. The cost and complexity of the regeneration system needs to be evaluated against other system requirements such as safety, reliability, maintainability, costs including procurement, maintenance, and operating costs, and customer desire for lower energy usage. Additional approaches for reducing energy consumption are discussed later in this book.

4.2 References

Shortley, G., and Williams, D. 1971. Elements of Physics, Fifth Edition. Prentice-Hall, Inc: Englewood Cliffs, NJ.

Beer, F. P., and Johnston Jr., E. R. 1976. Mechanics for Engineers, Dynamics Third Edition. McGraw-Hill Book Company: New York.

Chapter 5
Power Sources

Sometimes the initial requirement set specifies the power source, and other times this is left up to the engineer to decide. Most high-performance actuators are hydraulic, air, or electric because the underlying technologies are mature and they can offer the needed force and power. Other factors influencing power source selection include force and power density required at the actuation point, component availability and maturity, safety and inherent fail safe features, stiffness, control system and power supply maturity, and customer preference.

5.1 Hydraulic

Historically, most high-performance drives were hydraulic because it provides high power density, can incorporate hydraulic logic and control elements that do not need separate controls, and can be software or electrical hardware controlled through the use of electrohydraulic valves. As hydraulic actuators need a source of pressurized fluid, developing this hydraulic power unit (HPU) becomes an additional design task related to actuator design. While there can be individual HPUs for each actuator, it is more common to have a central HPU service several actuators. The design of the HPU then becomes dependent on the needs of the actuators for pressure, flow, cleanliness, and temperature control requirements. HPU flow rate and total fluid requirements can be determined from the operating cycle.

For example, it is important to understand these requirements so that the configuration of the HPU and component sizes can be accurately determined. It is a good idea to design some margin into the HPU to make up for unaccounted leakage, system inefficiencies, and growth. If the HPU is intended to supply all the pressurized fluid for the system, this additional capacity allows for future system enhancements without the need for replacing the HPU.

5.1.1 Hydraulic Symbols

Creating an easy to read, concise diagram of hydraulic systems is an important step in communicating and documenting the design. A popular method for creating diagrams is to use standard hydraulic symbols representing the various hydraulic components and their connections. While these symbols do not allow one to see the exact operation of components, they allow the design intent to be shown. Most symbols used today are based on ISO 1219. Symbols used in this book to illustrate hydraulic systems are shown in Appendix A. [MIL-HDBK-118 1993, A-11]

Some companies use pictorial representations of the components in their schematics. If done accurately, these provide insight into the exact working of valves and other components and are a great aid in troubleshooting problems. However, they are much harder to generate, take more time to understand, and cannot utilize standard symbols found in many design packages. In some cases, a combination of standard hydraulic symbols for standard components and pictorial representations for custom components can be an effective way for the engineer to communicate the design to others.

The choice of how to document a hydraulic system is ultimately the choice of the customer based on company standards and history. However, the engineer can offer suggestions to improve design efficiency and standardization.

5.1.2 Power Source

Choosing the power source for a hydraulic system can influence the feasibility of using hydraulics or some other approach such as electric. Power sources for a hydraulic system are generally either electric or engine driven. For mobile units, the most efficient choice is usually to drive the hydraulic pump directly from the engine. In this case, the engineer just needs to find a suitable pump mounting location on the engine where it can be driven to the appropriate speed and to get the required flow from the hydraulic oil tank.

It is possible to power an electric generator with the engine and then use an electric motor to drive the hydraulic pump. However, this approach is usually less efficient because there are power losses in the generator and electric motor in addition to the hydraulic pump losses. If the generator is 90% efficient, the electric motor is 95% efficient and the controller 98% efficient and then the system is only $.90 \times .95 \times .98 = 84\%$ efficient before it even gets to the pump. However, this needs to be viewed from a system viewpoint. If the engine can be run in a more efficient manner, when it drives the generator or the mechanical driveline to the pump is complex, then a generator/motor set may be more efficient. Of course adding a generator, an electric motor, and a controller can be additional cost items that may influence the decision. However, electrical needs are present in almost all systems now, so they may just need to be larger to drive a hydraulic pump.

The power source decision may also be directed by the requirements, customer, or user preferences. Even if the customer makes the decision on power source, the engineer should communicate the design impact of various power sources on the overall design so that the customer can make a fully informed decision. The engineer needs to keep current with all technologies influencing the power source and prime mover selection because new technologies can change the assessment of efficiency, cost, and future growth of the overall system.

5.1.3 Hydraulic Power Unit

The HPU for high-performance hydraulic actuators supplies pressurized fluid to the actuators and accepts low pressure from the return side of the actuators. A couple of different approaches for HPU designs are shown in the schematics of Figures 5.1 and 5.2.

Move from A to F

Elapsed Time (Seconds)

1.0 2.0 3.0 4.0 5.0 6.0 7.0 8.0 9.0 10.0 11.0

No.	Time	Function
1	0.50	Pick from A
2	0.50	Move from A
3	0.30	Rotate to B
4	0.50	Move to B
5	0.20	Exchange at B
6	4.00	Return to A
7	3.50	Move to C
8	2.50	Return to B
9	1.50	Rotate to D
10	0.50	Exchange at D
11	2.00	Return to C
12	2.00	Perform Operation at D
13	0.50	Pick from D
14	0.50	Move to E
15	0.50	Return to D
16	0.70	Place at F
17	0.40	Return to E

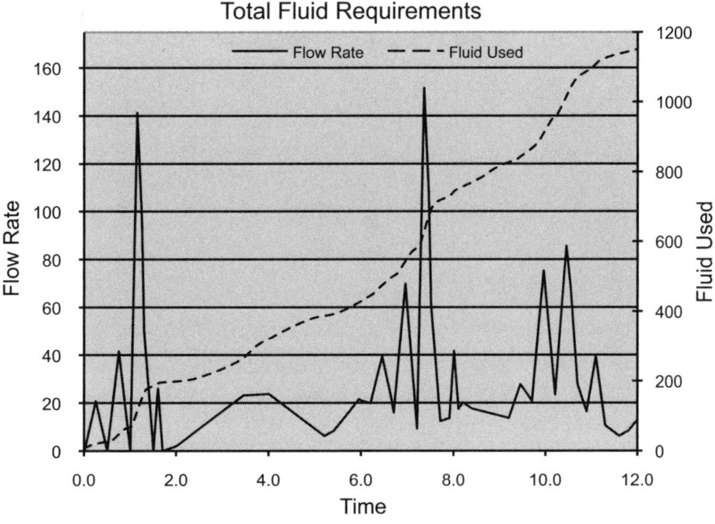

Figure 5.1 The HPU with fixed displacement pump (actuator pressure and return connections are on the left and the right loop represents an off-line filtration system).

An actuator needs to govern the design of the HPU. The fluid flow required from the HPU is determined from a fluid requirements analysis similar to the one shown in Section 5.1. The fluid requirements analysis is based on component force and/or torque requirements, operating duration, and operating frequency. These combine to determine the overall HPU power requirement. Other HPU components such as fluid, tank, pump(s), filter(s), accumulator(s), heat exchanger(s), and valves are selected based on the operating parameters and actuator component requirements. These are covered further in Section 5.1.4.

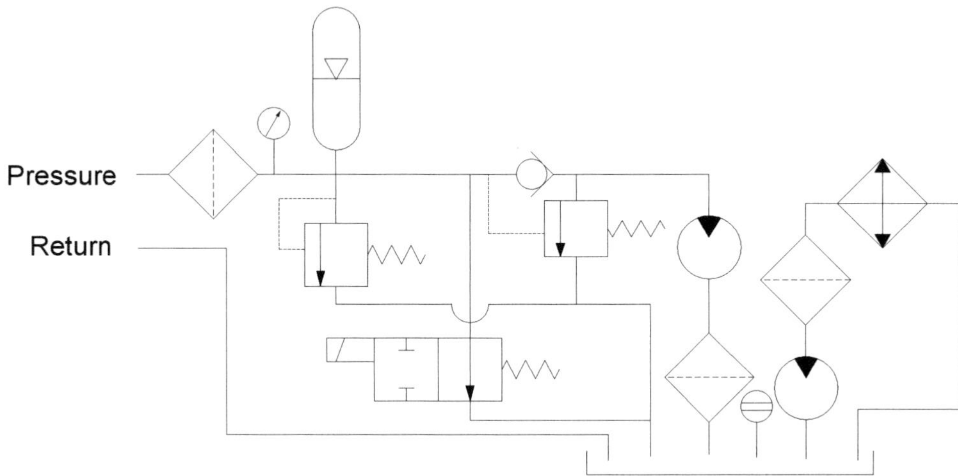

Figure 5.2 The HPU with variable displacement pump (actuator pressure and return connections are on the left and the right loop represents an off-line filtration system).

5.1.4 HPU Component Selection

HPU component selection is based on several factors that include organization or industry tradition and overall system operating requirements such as temperature range, environmental cleanliness, reliability, maintenance, and safety. One usually starts with the traditional factors as they represent the items that the company and/or customers are familiar with and stock. When these items are no longer available or cannot support new requirements, then the engineer must go through a selection process to define new components. Hopefully, a set of components can be identified that support the new requirements so if supplier issues arise, alternatives are available. The following sections discuss some of the items that need to be considered when designing and selecting HPU components.

5.1.5 Fluids

Unless the design is completely new, the first choice for a hydraulic fluid that should be a fluid currently used in other systems. However, fluid properties should be checked to make sure that they are compatible with new operating parameters and component requirements. Some properties include flammability, viscosity, temperature range, lubricity, cost, bulk modulus, and safety. Some fluids are flammable, especially when leaking under high pressure. As the pressurized fluid is released, it atomizes and when mixed with the oxygen in the air, it becomes highly flammable and dangerous. Flammability of hydraulic fluids is one of the reasons why some industries are replacing hydraulic actuators with other technologies. Viscosity of the fluid is also an important parameter to consider, especially how it changes over the operating temperature range. Viscosity at storage temperatures is usually not a big concern. The viscosity needs to support component operating needs at low temperatures, yet not be too thin to prevent component damage at high temperatures. Everyone familiar with hydraulic systems has experienced agonizingly slow operation at low temperatures and/or component damage because the viscosity became too low at high temperatures. Some hydraulic fluids are used to lubricate other components so their lubricity is an important consideration. However, the fluids and additives that provide good lubricity are not necessarily consistent with other properties desired in a hydraulic fluid. Therefore, this tradeoff may need to be made during fluid selection and accommodated during the design of other hydraulic components. As with everything else in a design, cost is an important consideration when selecting a hydraulic fluid. If the system volume requirements are low, then it may not seem like a big issue, but it could still be a significant percentage of the overall product's cost. It can also be a maintenance cost burden that the customer is sensitive about. As reviewed earlier, stiffness can play an important role in the performance of hydraulic actuators. The fluid bulk modulus (β) is a measure of how a fluid's volume changes with pressure and therefore directly affects a hydraulic component's spring rate. The engineer should also consider how susceptible a fluid is to retain air because any air entrained in the fluid greatly reduces its bulk modulus:

$$\beta = \frac{V_o \times (P_o - P_1)}{V_o - V_1} \tag{5.1}$$

where V_o = initial volume

V_1 = volume at Condition 1

P_o = initial pressure

P_1 = pressure at Condition 1

[The Lee Company 2000, M24]

Finally, overall safety of the fluid should be considered. Safety can include such areas as human toxicity and environmental impact. If the fluid is hazardous, extra care and features need to be incorporated into the design to minimize the chance of contact with it. Likewise, the fluid's environmental hazard level affects the precautions that must be incorporated into the design to minimize the chance that environmental contamination can occur. In addition to poor public relations, these factors can also greatly affect the overall cost of using a particular fluid.

5.1.6 Tank or Reservoir

The HPU tank is a rather simple component, mainly structural. As such, the engineer needs to determine the tank size, style, and mounting provisions. Tank size generally depends on total system volume, flow rate, and duty cycle. Essentially the tank needs to be big enough to give contamination a chance to settle out of the oil, air to separate from the oil, dissipate heat from the oil, and provide reserve capacity for any system leakage. Tank sizes generally range from 20% of the main pump flow rate for mobile equipment to 300% for stationary equipment. Given this range in tank size, determining the final tank size needs to be based on calculations for system leakage, volume change during operation, and time the oil spends in the tank. As it is desirable for oil to spend the maximum amount of time in the tank, return line connections should not be near the pump inlet. In addition, it is desirable to utilize baffles to force the oil to move and mix through the entire tank as it travels from the return connection to the pump. Regardless of the tank style, efficient designs utilize these baffles as structural tank elements. [Sperry Vickers 1967, 23–26]

Early in the design process, the engineer needs to decide if the tank style is open or closed from the environment. Closed tank designs can utilize a bladder in the tank or incorporate some sort of surge tank to limit fluid contamination. Closed tanks also offer the opportunity to utilize the entire tank volume when the system is on an angle or inverted, pressurize the entire hydraulic system that provides a precharge on the main pump minimizing cavitation problems. However, closed tanks are more complex and expensive as they need features to allow changes in oil volume through the entire temperature range, changes in actuator oil volume requirements, and fluid loss through leakage. These volume changes can be significant so that the entire hydraulic system volume needs to be assessed to make sure the tank design can accommodate the entire volume change. Closed tanks also need to incorporate a filling and draining approach that is acceptable to the customer. Open tank designs are much simpler. They just need to hold the required fluid volume and allow for fluid expansion. A simple vent or breather can allow the volume change while limiting the amount of contamination that enters the system. Whether the HPU is a stand-alone unit or more fully integrated into the system structure, mounting provisions for all HPU components need to be accommodated.

Before mounting provisions can be determined, all components of the HPU need to be identified and selected. Not all components need to be mounted directly to the HPU, but it is usually a convenient place to mount them and keeps most hydraulic maintenance actions in one area. When determining mounting locations, the engineer should minimize plumbing connections and bends so that there are fewer flow losses and leakage points. Design of mounting provisions should also accommodate dynamic loads from operation including vehicle motion if it is a mobile system, loads from personnel grabbing, standing, or otherwise using the equipment to move around the system, and consider induced vibration or shock loads.

5.1.7 Pumps

The HPU may just have a single main pump or may need several pumps to provide reliable, efficient operation. Generally, there is a main pump to provide pressurized fluid at a flow rate sufficient to operate the hydraulic components as desired. If the operational duty cycle is low or extremely high flow rates are required, an accumulator may be used. Accumulators can even out flow requirements so that the main pump needs only to provide the average flow rate for the duty cycle. If the main pump has variable displacement, then it needs only to provide the flow required by the system and consumes little energy when no flow is required. A popular type of variable displacement pump is an axial piston pump where the piston stroke is governed by the position of the swash plate. When the swash plate is perpendicular to the piston axes, the pistons do not reciprocate and not fluid is pumped. As the swash plate angle moves from perpendicular, the piston stroke is lengthened—pumping more fluid. A fixed displacement pump always pumps the same amount of fluid for each revolution of the pump. When connected to a constant speed drive, it provides constant flow to the system. If the system does not require fluid flow, an unloading valve in the system can remove the load from the pump by allowing the pump to pass the fluid directly back to the tank without pumping it to full system pressure. This minimizes the energy required and heat generated. Other pumps in the system may include charge pump, offline loop pump, and auxiliary pump. These tend to be fixed displacement pumps because they can perform the required function(s) in a small, reliable, low cost package.

A charge pump supplies low pressure fluid to the main pump, minimizing the possibility of main pump cavitation. This can be a serious problem for variable displacement pumps that are located at a long distance from the tank. When the system demands full flow, the pump goes to full stroke immediately and the fluid in the suction (inlet) line cannot accelerate fast enough to keep the pump supplied. Cavitation occurs when the pressure in the supply line drops below the fluid's vapor pressure forming vapor bubbles. These bubblers can then rapidly collapse, creating a shock wave when the fluid gets in an area of the system that exceeds the vapor pressure. This process can rapidly damage hydraulic pumps and other equipment, so it should be avoided. The charge pump should be sized to provide more flow than the main pump, with excess flow diverting back to the main tank. [Sperry Vickers 1967, 37–60], [John Deere 1979, 2-1 to 2-20]

An offline pump generally sends a constant stream of fluid to a filter and/or heat exchanger. Both these components are more reliable and operate with greater efficiency when they are used at a constant flow and not subjected to the rapid flow changes that may occur in the main flow circuit. The offline pump size needs to be selected so that it provides the flow necessary to achieve the filtration needed or adequately conditions of the fluid temperature.

Sometimes an auxiliary pump is added to an HPU to provide limited operation when the main pump is not working or if full flow from the main pump is not desirable. The auxiliary pump should be more tolerant of fluid contamination than the main pump, in case contamination caused main pump failure or failure of the main pump added contamination to the system. The auxiliary pump may be used for maintenance actions where personnel may be exposed to moving machinery. In this case, it is desirable for the pump to operate at a pressure level and a flow rate that offers a safe environment for the personnel.

5.1.8 Prime Mover

Selection of the prime mover for the HPU is often based on the most convenient source. For industrial settings, this is usually an electric motor. For mobile applications, it is usually a dedicated engine or power takeoff on an engine also used for other functions. Other options include generating electric power on mobile equipment that can be used to power an electric motor driving the hydraulic pump(s). While this latter option is usually less efficient overall, it offers the option to power multiple pumps and/or mount them remotely where they can access the rest of the hydraulic components better. This convenience may offer certain packaging and reliability improvements but can increase system cost and decrease overall efficiency.

5.1.9 Filters

The fluid cleanliness level required is typically driven by the actuator components selected. The engineer needs to follow the component manufacturer's recommendations and design the system for the most sensitive component. A widely accepted fluid cleanliness standard is ISO 4406/SAE J1165, which gives three range numbers for cleanliness according to particle size. The first number is for particles greater than 4 mm, the second is for particles greater than 6 mm, and the third number is for particles greater than 14 mm. For sensitive components operating at high pressure, such as servo valves, the rating may be in the range of 16/12/10 to 19/16/13. Sometimes, a manufacturer specifies two values, a larger one for function and a smaller one to minimize wear. Other less sensitive components such as cylinders and flow control valves may have a rating up to 21/18/15. Another fluid cleanliness standard sometimes used is National Aerospace Standard (NAS) 1638. This standard has cleanliness classes ranging from 00 to 12, with 00 being the cleanest and 12 the dirtiest. As new, unused oil can have an ISO 4406 cleanliness of 21/18/15, it needs to be filtered before it is used with sensitive components. [Hydraulics & Pneumatics 2015, 1]

There are several ways to filter a hydraulic system, but the best is to use an offline filtration circuit along with a pressure line filter between the pump and sensitive components. In some cases, where designers do not want to add the additional components to support an offline filtration circuit, a return line filter is added. To protect the pump from ingesting large particles from the tank, it is also a good idea to utilize a coarse filter or strainer at the pump inlet. Each of these filters needs to be selected to perform the selected function. Filtration components at the pump inlet need to be sized to make sure that they can handle the maximum pump flow rate without causing the pump to cavitate. If the pump inlet line is long or small, it is better to delete the pump inlet filter than guarantee pump failure through cavitation. Filter performance is often captured as a beta ratio, which is the ratio of upstream particles to the downstream particles. For example, if the upstream particle count for 10 mm and larger particle count is 100,000 and the downstream count is 25,000, the beta ratio is 4:

$$B_x = \frac{\text{Upstream\_Particles}}{\text{Downstream\_Particles}}$$
$$B_{10} = \frac{100{,}000}{25{,}000} \tag{5.2}$$
$$B_{10} = 4$$

The filter's efficiency is then

$$\text{Eff}_x = \left(1 - \frac{1}{B_x}\right) \times 100$$
$$\text{Eff}_{10} = \left(1 - \frac{1}{4}\right) \times 100 \tag{5.3}$$
$$\text{Eff}_{10} = 75\%.$$

Selecting filters for pressure lines, return lines, and/or offline systems is best done with the help of the filter manufacturer because the desired cleanliness level can be obtained several ways, with some combination of filter efficiency, flow, and contamination ingestion rate. High-performance systems usually subject pressure and return filters to significant flow surges, so pressure and return filters do not work as efficiently as filters used in an offline circuit. Some manufacturers have filter selection software available to help customers select the correct filter from their product line. However, one can also get a rough idea of filtration requirements from the initial fluid cleanliness, the contamination ingestion rate, total system fluid volume, flow rate, and filter efficiency.

5.1.10 Accumulators

Using and sizing an accumulator depends on pump flow rate, pump response, and actuator flow requirements. However, the ability to easily store energy in an accumulator is one of the big advantages of hydraulic actuators over other types of actuators. They can also be a limited source of backup power if the pump or prime mover fails. Based on fluid requirements, such as those shown in Figure 5.3 and the pump flow rate, the required accumulator fluid volume can be calculated. The design needs to make sure that the accumulator maintains adequate operating pressure through the entire cycle. Standard accumulators are designated by the gas volume they can hold when no fluid is present. For example, a 18.9-L (5-gallon) accumulator has 18.9 L (5 gallons) or 18,927 cm^3 (1155 in^3) of gas volume. Calculating the system pressure is as simple as utilizing the ideal gas law for reversible adiabatic expansion and compression of a gas:

$$P_1 V_1^n = P_2 V_2^n \tag{5.4}$$

where P_1 = accumulator gas precharge pressure

V_1 = accumulator volume

P_2 = system pressure with V_3 fluid in the accumulator

$V_2 = V_1 - V_3$

V_3 = volume of fluid in the accumulator

n = gas constant

[Shortley and Williams 1971, 337–338, 385–389], [Streeter and Wylie 1975, 670–672]

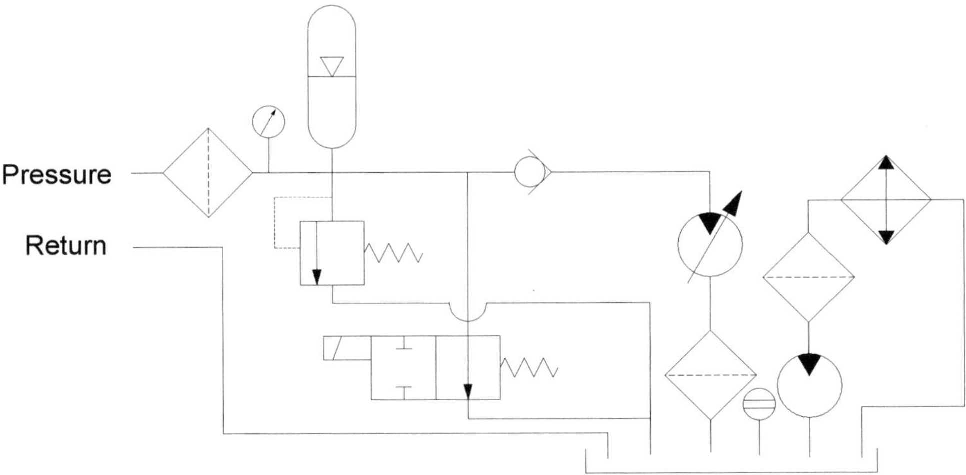

Figure 5.3 Fluid requirements.

If the system is operating slowly or is at rest for extended periods of time the gas constant, n can approach 1. Other times, when it is operating rapidly, n is usually 1.3–1.4 for a nitrogen charged accumulator. [The Lee Company 2000, M54] The engineer needs to assess all the operating scenarios to make sure that the system is going to work satisfactorily all the time. If necessary, separate gas bottles can be added to the system to increase gas volume or custom accumulators can be designed and procured from suppliers. Sometimes, accumulators are added or utilized to manage pressure spikes that occur from rapid changes is fluid flow in either the pressure or return lines. After determining the accumulator volume required, the engineer needs to decide on the type of accumulator to use.

There are four basic types of accumulators: bag, diaphragm, metal bellows, and piston. Each of these isolates the fluid from the environment or charging gas. While an accumulator can be charged with a spring, weight, or gas, most high-performance systems utilize a gas, typically nitrogen, because it is more compact and light for the force developed. Bag accumulators store the gas in a bag that expands like a balloon pressurizing the fluid. Their usual failure mode is rupturing of the bag, which then allows the gas to mix with the fluid. Diaphragm accumulators are much like a bag accumulator except that instead of a full bag for the gas, a diaphragm attached to the container walls separates the gas and fluid. The usual failure mode is again rupturing the diaphragm, allowing the gas to enter and mix with the fluid. The metal bellows is really a version of the bag accumulator where a metal bellow is used instead of an elastomer bag to separate the gas from the fluid. While metal bellows accumulators tend to be expensive, they are also very reliable and compact, so there is a tradeoff between cost and reliability requirements. Finally, a piston accumulator utilizes a piston to separate the gas and fluid. They tend to be reliable and fail gracefully but have more friction losses than the other types. While fluid level sensor can be integrated into each accumulator type, it is usually easiest to integrate them into piston accumulators. These sensors can verify the fluid level for control and/or safety reasons.

Accumulators can be located in any location of a hydraulic system, but they are usually located on the pressure side of the system just downstream from the pump. This location allows the accumulator(s) to service all actuators and the close proximity to the tank allows for convenient discharging of all charge fluid when the system is shut down. It is important to make sure that an accumulator is fully discharged when the system is shut down to prevent any safety issues associated with unintended motion or pressurized fluid when the system is expected to be "safe" or maintenance is performed.

5.1.11 Heat Exchangers

Unfortunately, high-performance hydraulic systems tend to be somewhat inefficient because there are high accelerations, velocities, and decelerations. As all this high-energy fluid passes through the values and other system components, flow losses end up as heat. For example, if a move takes 1000 cc (61 in^3) of 20.7 MPa (3000) psi fluid to move a 45-kg (100-lb) payload horizontally 305 mm (1 ft) and vertically 305 mm (1 ft). If the horizontal coefficient of friction is .1, the work done is

$$W = .1 \times 305/1000 \times 45 \times 9.806 + 305/1000 \times 45 \times 9.806$$
$$W = 148 \text{ Nm (110 ft-lb)}.$$

The energy consumed is

$$E = 20.7 \times 1000000 \times 1000/1000000$$
$$E = 20{,}700 \text{ Nm } (15{,}268 \text{ lb-ft}).$$

The heat generated in the fluid is

$$H = 20{,}700 - 148$$
$$H = 20{,}552 \text{ Nm } (15{,}158 \text{ lb-ft})$$
$$H = 15{,}158 \text{ lb-ft} \times \frac{1 \text{ BTU}}{778 \text{ lb-ft}}$$
$$H = 20.552 \text{ kJ } (19.5 \text{ BTU}).$$

For this system, 8339 J (7.89 BTU's) must be dissipated for every cycle. If the system has 50 L (13 gallons) of hydraulic fluid with a density of 830 kg/m^3 (.03 lb/in^3) and a specific heat capacity of 2.09 kJ/kg°C (.5 BTU/lb-°F), the temperature rise per cycle is

$$\Delta T = 20.552\ \text{kJ} \times \frac{1}{50\ \text{L}} \times \frac{\text{L}}{0.001\ \text{m}^3} \times \frac{\text{m}^3}{830\ \text{kg}} \times \frac{\text{kg °C}}{2.09\ \text{kJ}}$$

$$\Delta T = 0.24\text{°C}\ (0.43\text{°F}).$$

One can see that the fluid temperature can quickly become too hot if there is not thermal management. If this system operates at 100 cycles/min, 2055 kJ need to be dissipated each minute for the system to remain at a constant temperature. [The Lee Company 2000, M20] While this is a somewhat crude analysis, it illustrates the need for a heat exchanger to manage the fluid temperature. From the fluid's viewpoint, it does not matter what type of heat exchanger is used as long as it removes enough heat from the system.

The most common heat exchangers used are fluid to air and fluid to water. Selection of the type is usually based on the media and space available to dissipate the energy. Unless the engineer has the ability and means to design a custom heat exchanger, the heat exchanger should be selected after contacting manufacturer to make sure it has adequate capacity and can withstand the operating and storage environments.

5.1.12 General Hydraulic Power Unit Valves

Valves for the HPU are generally simple because they do not have to precisely control flow. Valve use includes tank fill and drain, accumulator dump control, offline loop control, and main supply control. If flow rates are small, there are many types of valves to choose from. High flow rates require more thought and design effort. The engineer also needs to decide if the valves can be operated manually or need remote or automatic control.

Flow control valve types generally include ball, poppet, and spool, with manual, remote, or solenoid control. Ball valves are generally manually actuated and support high flows, but may be hard to operate at high pressures. Poppet valves are generally remotely actuated and can support various flow rates, including high flows. Some of the most convenient large poppet valves conform to DIN 24342, which support flows to 11,350 L/min (3000 gpm) at a pressure drop of around 1 MPa (150 psi). These can be controlled remotely through a solenoid actuated valve and can be configured various ways to suit the circuit needs. Spool valves can be operated manually, remotely, or by a solenoid. If the valve is small, it can be operated directly by a solenoid, and large valves usually have two or three stages before the valve gets large enough to support the required flow rate. Porting and valve lands can be configured to achieve the flow control desired. Each of these valves can be custom made or selected from catalogs.

Other valves in the HPU circuit may include check valves to make sure that flow does not drain out of the greater system when the HPU is shut down. Pressure relief valves to make sure that system pressure never exceeds system design pressure if there is a blockage or pump control failure. Finally, if the system utilizes a fixed displacement pump, an unloading valve may be used to reduce system pumping pressure when pump flow is not required.

A simple way to create a custom flow circuit for the HPU is to use cartridge valve bodies in a custom manifold. This is a low-cost and risk approach because the valve components have been already designed, tested, and fabricated. The only risk is in the design of the manifold, but standard cartridge cavity tools are available to make the cavities and porting between cartridges and exterior connections are drilled holes in the manifold. One just needs to make sure that the pressure drop through the valves and manifold does not exceed design requirements at maximum flow.

If one needs to design custom valves, the first two items that need to be determined is valve function and flow. Valve function should be derived from the schematic and performance parameters. While careful design and analysis during valve design increases the probability that the valve will function as expected when manufactured, design experience also helps. There are many subtleties in component selection and design that impact flow through a valve that are often not recognized by a new designer until the valve is under test. Additional design details are presented later in this book.

5.1.13 High-Performance Actuator Control Valves

Control of a high-performance actuator with a hydraulic valve requires close coupling of the valve to the actuator to minimize compliance in the lines between the valve and actuator and high valve response rates. The type of actuator and level of control built into the actuator defines the sophistication required from the control valve. For systems with load-controlled acceleration and internal deceleration features such as shown in Figures 5.27–5.29, a simple high response ON/OFF valve can provide actuator start, OFF, and reverse functions. If the actuator does not have defined end points or is designed to follow a specific velocity profile such as those shown in Figures 3.1–3.4, a more sophisticated control valve is required. Depending on the flow rate required to meet the power requirements, these valves may require one, two, or even three stages. To ensure system, equipment, and personnel safety, the overall actuator control scheme needs to be assessed so that the actuator operates only when commanded, there is a method to bring it to a controlled stop if there is a system failure and it can safely hold the load when stopped. Many of the flow control valves described in Section 5.1.12 can be utilized for simple ON/OFF control, emergency stop, and pressure control functions. However, sophisticated actuator is generally accomplished with high-performance electrohydraulic servo valves, such as those shown in Figure 5.4. If a somewhat lower performance level can be tolerated then proportional valves such as those shown in Figure 5.5 may be used. [Rexroth 1984, 199, 252–254], [John Deere 1979, 3-1 to 3-22]

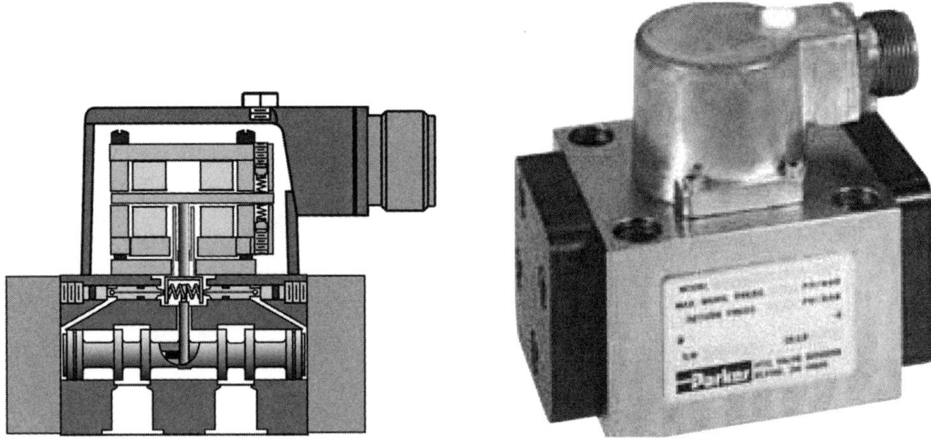

Figure 5.4 The two-stage servovalve. Parker DY15 Series Servovalve (www.parker.com). Image courtesy of Parker Hannifin Hydraulic Valve Division.

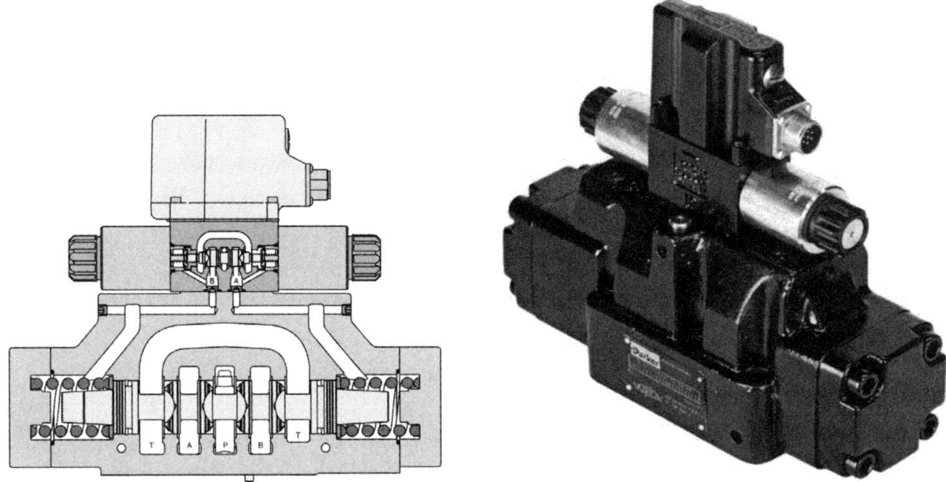

Figure 5.5 The proportional valve—the Parker Series D*1FB pilot operated proportional valve (www.parker.com). Images courtesy of Parker Hannifin Hydraulic Valve Division.

5.1.14 References

Shortley, G., and Williams, D. 1971. Elements of Physics, Fifth Edition. Prentice-Hall, Inc: Englewood Cliffs, NJ.

John Deere Service Publications. 1979. Hydraulics, Fundamentals of Service. Deere & Company: Moline, IL.

Sperry Vickers. 1967. Mobile Hydraulics Manual. Sperry Rand Corporation: Troy, MI.

The Lee Company Technical Center. 2000. Technical Hydraulic Handbook. The Lee Company: Westbrook, CT.

Rexroth Worldwide Hydraulics. 1984. Components for Hydraulic Proportional and Servo Systems, Electronics and Accessories. The Rexroth Corporation: Bethlehem, PA.

Hydraulics & Pneumatics. 2015. Hooked on Hydraulics Fluid Cleanliness. Penton Media, Inc: Overland Park, KS.

Streeter, V. L., and Wylie, E. B. 1975. Fluid Mechanics. McGraw-Hill Book Company: New York, NY.

U.S. Department of Defense. 1993. MIL-HDBK-118, Design Guide for Military Applications of Hydraulic Fluids. U.S. Department of Defense: Fort Belvoir, VA.

5.2 Pneumatic Systems

Design of pneumatic systems is similar to hydraulic systems except that air is significantly more compressible than hydraulic fluids. As compressibility of hydraulic fluids can complicate the design of high-performance hydraulic systems, one can see that compressibility is even more significant in the design of pneumatic systems. In fact, this is one of the reasons pneumatic systems are not often used for high-performance drives. One way to characterize the stiffness of a fluid is by its bulk modulus. For hydraulic fluids, the bulk modulus can be around 1378 MPa (200,000 psi), depending on how much air is entrained in the fluid. The effect of bulk modulus can be represented by the following equation:

$$\beta = \frac{V_1 \times (P_2 - P_1)}{V_1 - V_2} = \frac{\rho_1 \times (P_1 - P_2)}{\rho_2 - \rho_1} \tag{5.5}$$

$$\beta = \frac{\rho_1 dP}{d\rho} \tag{5.6}$$

where P_1 = initial pressure

V_1 = initial volume

P_2 = final pressure

V_2 = final volume

ρ_1 = initial density

ρ_2 = final density

dP = change in pressure

$d\rho$ = change in density

β = bulk modulus

For ideal gases

$$P_1 V_1^n = P_2 V_2^n \tag{5.7}$$

From basic thermodynamics, we know

$$PV = mRT \tag{5.8}$$

At constant temperature (isothermal):

$$\frac{PV}{m} = RT = \text{Constant (C)} \tag{5.9}$$

or

$$\rho = \frac{P}{C} \tag{5.10}$$

[Shortley and Williams 1971, 338, 388]

Substituting (5.10) into (5.5)

$$\beta = \frac{\frac{P_1}{C} \times (P_2 - P_1)}{\frac{P_2}{c} - \frac{P_1}{c}} \tag{5.11}$$

$$\beta = P_1 . \tag{5.12}$$

At an operating pressure of 20 MPa (2900 psi), the isothermal bulk modulus is 20 MPa or about 1.5% of a typical hydraulic fluid. If the system is operating rapidly in an isentropic process, then temperature is not constant, but the system is at constant entropy at which case n of (5.7) is not 1, so

$$\rho^n = \frac{P}{C} \tag{5.13}$$

or

$$P = C\rho^n \tag{5.14}$$

$$\frac{dP}{d\rho} = nC\rho^{n-1} . \tag{5.15}$$

[Streeter and Wylie 1975, 333–340], [Wark 1977, 80–88], [The Lee Company 2000, M54]

Substituting in (5.6)

$$\beta = \rho nC\rho^{n-1} \tag{5.16}$$

$$\beta = nC\rho^n . \tag{5.17}$$

Knowing from (5.14) and substituting in (5.17)

$$\beta = n\rho . \tag{5.18}$$

If n = 1.4 for an isentropic application

$$\beta = 1.4 \times 20 = 28 \text{ MPa}$$

or about 2% of a typical hydraulic fluid. This low fluid stiffness is one of the primary reasons that pneumatics are not used for large high-performance systems. If power requirements are low and pneumatics can provide a convenient power source because no fluid needs to be returned to the tank, any leaks do not cause any sort of pollution or health hazard. However, the designer needs to be aware of the safety implications that accompany a gas under pressure. Due to their compressibility, gases can store significant energy when compressed, so the designer needs to make sure that processes for personnel and equipment safety are in place. In simple terms, the amount of energy stored in pneumatic and hydraulic systems as follows:

$$\begin{aligned} E &= \frac{1}{2}k(\Delta x)^2 \\ k &= \frac{\Delta F}{\Delta x} \\ \beta &= \frac{V \times \Delta P}{\Delta V}. \end{aligned} \tag{5.19}$$

For a cylinder

$$V = A \times x$$

so

$$\beta = \frac{A \times x \times \Delta P}{A \times \Delta x} \tag{5.20}$$

or

$$\Delta x = \frac{x \times \Delta P}{\beta} \tag{5.21}$$

and

$$k = \frac{A \times \Delta P}{\Delta x} \tag{5.22}$$

so

$$\begin{aligned} E &= \frac{1}{2} \times \frac{A \times \Delta P}{\frac{x \times \Delta P}{\beta}} \times \left[\frac{x \times \Delta P}{\beta}\right]^2 \\ E &= \frac{1}{2} \times A \times \Delta P \times \left[\frac{x \times \Delta P}{\beta}\right] \\ E &= \frac{A \times x \times \Delta P^2}{2 \times \beta}. \end{aligned} \tag{5.23}$$

For two identical systems, except with a different bulk modulus, the energy ratio is

$$\frac{E_1}{E_2} = \frac{\beta_2}{\beta_1}. \tag{5.24}$$

If system 1 is pneumatic with $\beta = 28$ and system 2 is hydraulic with $\beta = 1378$, we see that the pneumatic system has 50 times more stored energy. This extra stored energy is what can make a pneumatic system much more dangerous than a hydraulic system if there is a component failure. The energy can propel components at high velocities in unpredictable directions that can damage other hardware and injure or kill personnel.

However, sometimes a pneumatic power source is available from industrial or mobile systems, so it is desirable to utilize it. Also, it can be clean and does not pose a combustible threat for actuators used in a combustible environment. If pneumatics are selected to power a high-performance actuator, then the design can be approached the same as a hydraulic system except the extra compressibility needs to be accounted for and there is no need for a return line because the low pressure side of the actuator can be vented to atmosphere.

5.2.1 References

Shortley, G., and Williams, D. 1971. Elements of Physics, Fifth Edition. Prentice-Hall, Inc: Englewood Cliffs, NJ.

The Lee Company Technical Center. 2000. Technical Hydraulic Handbook. The Lee Company: Westbrook, CT.

Streeter, V. L., and Wylie, E. B. 1975. Fluid Mechanics. McGraw-Hill Book Company: New York, NY.

Wark, Kenneth. 1977. Thermodynamics. McGraw-Hill Book Company: New York, NY.

5.3 Electric

Electric drives are capturing a significant portion of the high-performance drive application as electric drive technology matures and component suppliers develop smaller, lighter, and more efficient electric drive components along with supporting power electronics and electronic control systems. As a review for electric drives, the following section offers a quick summary of electric circuit basics. These include voltage, current, resistance, capacitance, and inductance and how they relate to each other.

Voltage is analogous to pressure in a hydraulic system because it is the electromotive force that produces current flow in an electrical conductor. Current is analogous to the flow rate in a hydraulic system because it is the flow rate of an electrical charge in a conductor. They are related to each other through resistance which is the change in voltage divided by the change in current as represented by Ohm's law shown in

$$V = iR \tag{5.25}$$

where V = voltage, i = current, and R = resistance.

Capacitance relates to stored energy because it is the change in electric charge divided by the change in voltage. In function, it is similar to an accumulator in a hydraulic system

$$C = \frac{Q}{V} \tag{5.26}$$

where C = capacitance, Q = electric charge, V = voltage, and i = current.

Inductance is the change in voltage caused by a charge flowing through a conductor

$$L = \frac{V}{i/s} \tag{5.27}$$

where L = inductance, V = voltage, and i/s = current per second.

The work done by an electric drive depends on the amount of electrical energy that is converted to mechanical work and how much is stored in the magnetic field:

$$E_{\text{in}} = E_{\text{Mech Work}} + E_{\text{Magnetic}} .$$

An electric drive can be thought of as an input signal converted to an output motion through an input circuit, a distribution circuit, and an excitation circuit. The input circuit takes the input signal or command and translates it to a form that can be used by the distribution circuit. Characteristics translated by the input circuit may include pulse height, width, rise time, and fall time as defined by the desired drive performance (Section 3.1). The distribution circuit takes the information from the input circuit and generates the excitation signal that can be sent to the excitation circuit. The excitation circuit receives the signal from the distribution circuit and amplifies it to the level needed to move the electric drive as commanded. The excitation circuit usually consists of power transistors that can supply the current needed at the drive operating voltage to overcome drive inductance and move the actuator. The excitation circuit needs to switch the drive current ON at the appropriate time and make sure that it rises as needed and then reduce the current in the drive windings at the end of the excitation. At this point, it also needs to reduce the voltage at the transistor collector from any electromotive force generated by the drive windings. Some methods to speed up current rise and fall include series resistance, constant-current, and high-voltage/low-voltage switching. A block diagram of these functions is shown in Figure 5.6. This sequence of events is shown in Figure 5.6.

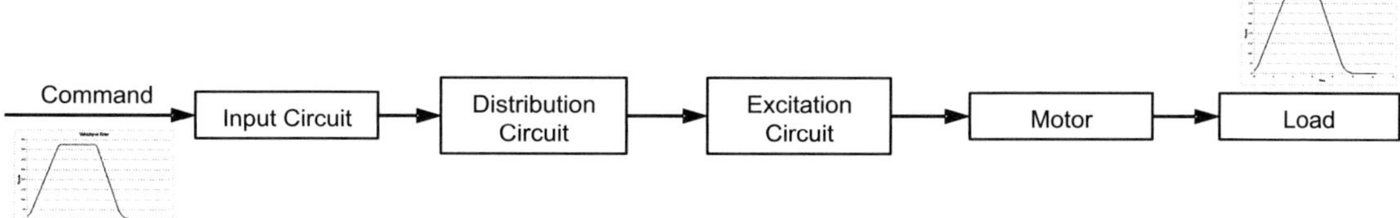

Figure 5.6 The electric drive.

The torque developed by an electric servo motor is a function of the current and air gap flux as shown in

$$T = K'i_a\gamma \tag{5.28}$$

where T = torque, K' = constant, i_a= armature current, Υ = air gap flux, $\gamma = K_f i_f$, K_f = field constant, and I_f = field current.

If the field current becomes constant, then (5.28) becomes

$$T = Ki_a$$

where T = torque, K = motor torque constant, and i_a = armature current.

When a motor is operating, a voltage proportional to the product of the flux and angular velocity is produced. At a constant flux, this voltage is directly proportional to the angular velocity of the motor and is called the back emf voltage

$$V_b = K_b\omega \tag{5.29}$$

where V_b = back emf voltage, K_b = back emf constant, and ω = motor angular velocity

For an armature controlled dc servomotor, speed is controlled by the armature voltage. The general equation for the armature circuit is shown in

$$V_a = L_a\frac{d_{i_a}}{dt} + R_a i_a + V_b \tag{5.30}$$

where V_a = armature voltage, R_b = armature resistance, i_a = armature current, V_b = back emf voltage, and $\frac{d_{i_a}}{dt}$ = derivative of armature voltage with respect to time.

If L_a is small, then

$$V_a = R_a i_a + V_b .$$

Combining with (5.29) gives

$$V_a = R_a i_a + K_b\omega$$

or the maximum theoretical motor speed

$$\omega = \frac{V_a - R_a i_a}{K_b} . \tag{5.31}$$

[Shortley and Williams 1971, 514, 532, 659], [Ogata 1992, 215–217]

Although very large mining equipment has utilized electric drives for many years, the drives did not provide the rapid, precise movements typically seen in modern high-performance actuators. The maturing of brushless dc motors, controllers, and power supplies made electric drives feasible where people had previously used hydraulic actuators. Using electric drives requires a close working relationship between mechanical engineers who create and specify the mechanical components, electrical engineers who create and specify the drive controllers and power supplies, and control engineers who have to make the system work. An organization that has access to this talent has the ability to successfully design and field high-performance actuators with force and power ranges within the current component state of art. Force and power limits are continually increasing for these drives, so they now successfully compete with hydraulic actuators in many applications. While people are cautious about the safety associated with the high-voltage, high-current requirements of these high power drives, they also appreciate the lack of hydraulic fluid that has leakage, fire, temperature, and environmental concerns.

High-performance electric drives tend to require the skills and knowledge found in mechanical engineers, electrical engineers, and control engineers. Depending on the organization and the level of drive customization, the design may involve the efforts of an engineer that does all three or a group of engineers with each individual being highly skilled in their area of specialty. As this book addresses the mechanical design of high-performance actuators, the electrical and control engineering portions of electric drive development, including power sources and power supplies are not covered in detail. Discussion on electric motor selection is covered in Section 5.4.1.4.

5.3.1 References

Shortley, G., and Williams, D. 1971. Elements of Physics, Fifth Edition.

Ogata, K. 1992. System Dynamics. Prentice Hall: Englewood Cliffs, NJ.

5.4 Actuator Detailed Design

5.4.1 Motor Selection

If a motor is utilized as the prime mover within the actuator, the motor selection process needs consider performance, duty cycle, available energy sources, thermal effects, and user serviceability. These can be captured in a set of motor requirements, which then lead the design toward an electric motor, a hydraulic motor, or a pneumatic motor. Once the power source has been selected, the actuator engineer can proceed with final design and selection of the motor and remaining components so that the entire actuator package satisfies all requirements. In general, the designer needs to carefully examine the design concept to understand the loads being imposed on the motor before proceeding. These loads should be captured in a set of motor requirements, so that they can be communicated to the motor supplier. Some general items that should be considered for all motors are presented in the following sections.

5.4.1.1 Motor Shaft Loads

Shaft loads are made up of axial and radial loads. Axial loads can come during installation when the motor is assembled and retained to a fixed shaft or if the shaft is not fixed and loads from the shaft or screw are transferred to the motor. Most motors are not designed to handle significant axial loads, so these loads need to be compared with the axial load capacity of the motor. Radial loads can come from assembling a motor into a fixed shaft where radial misalignment or runout causes the shaft to be loaded radially as it rotates or from drive loads transferred directly to the motor shaft. As with axial loads, most motors are not designed to accommodate significant radial loads. Shaft loads can be managed through the use of couplings and joints that limit axial and or radial loads. When incorporating these components, the design engineer needs to accommodate their space claim, alignment requirements, and stiffness/backlash characteristics in the design layout and performance calculations. When the shaft connection approach has been defined, the engineer needs to make sure that the standard motor can accommodate the axial and/or radial loads, provide axillary bearings to accommodate the loads, or have the motor customized with bearings, shafting, and housings that can accommodate the loads.

5.4.1.2 Motor Torque Requirements

The torque capacity of a motor needs to move, accelerate, and decelerate the load as desired. Adequate torque can be easily overlooked if the designer just concentrates on power, but it is essential to have adequate torque to achieve the performance desired. Motor torque needs to meet, with some margin, all actuator torque requirements such as acceleration, deceleration, unbalance, externally induced loads, and emergency stopping.

To calculate the required motor torque, the engineer needs to determine the static and dynamic loads on the drive. As discussed earlier, these loads include gravity-induced loads, acceleration loads, and friction loads.

Gravity loads are torque or unbalance loads that follow Newton's first law of motion, $F = ma$ where m is the mass of the load and all hardware between the load and motor and a is the gravitational acceleration. The load at the motor is adjusted by the speed ratio between the unbalance load and the motor. If motor speed is different from the load speed, the unbalance load is adjusted by the speed ratio, or

$$\text{Motor Unbalance Load} = \text{Unbalance Load} \times \frac{\text{Load Speed}}{\text{Motor Speed}} \tag{5.32}$$

Other components between the load and motor that affect unbalance also need be accounted for. These components can add or subtract from the load unbalance. Sometimes additional mass is added to the system to counterbalance or equilibrate the load. Whatever the case, unbalance effects on the motor need to be calculated to adjust for the total motor unbalance load. Equation 5.32 can be used to calculate each component's contribution to the total motor unbalance load by determining component unbalance and multiplying it by the speed ratio between the component and motor. One needs to make sure that they follow an established sign convention, such as clockwise unbalance being positive (+) and counter clockwise unbalance being negative (–).

$$\text{Component Unbalance at the Motor} = \text{Component Unbalance} \times \frac{\text{Component Speed}}{\text{Motor Speed}}. \tag{5.33}$$

The engineer should also make sure that any unbalance changes due to changes in mechanism geometry as the actuator goes through its entire motion are also considered. In some cases, geometry changes can make the unbalance load on the motor change directions. While such a mechanism may minimize motor unbalance loads, it may cause overall actuator precision and control issues because any backlash in the drive shows up at the load when the unbalance changes direction. Motor torque to manage unbalance loads should remain in the continuous operating range of the motor; the continuous rate torque line is shown in Figure 5.7 or 5.8.

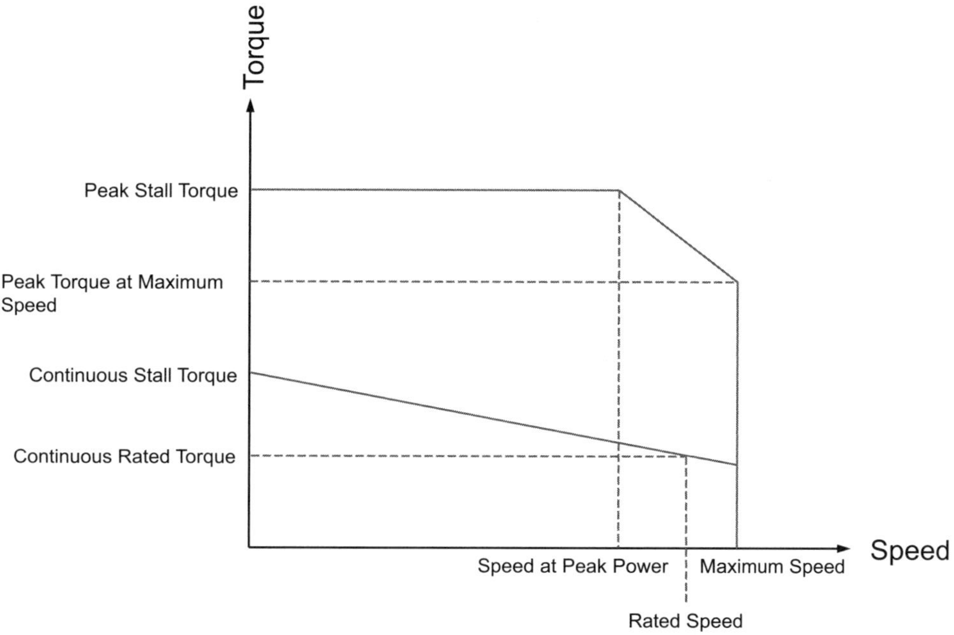

Figure 5.7 Typical electric motor torque/speed characteristics.

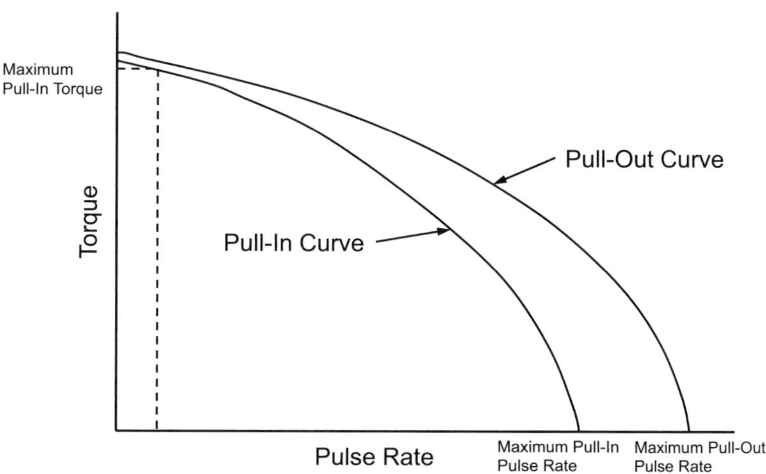

Figure 5.8 Stepper motor torque—versus pulse rate.

Acceleration loads are torque or unbalance loads that also follow Newton's first law of motion, $F = ma$, for linear motion and $T = I\alpha$, where I is the mass moment of inertia for the load and all hardware between the load and motor and α is the rotational acceleration associated with *I*. [Shortley and Williams 1971, 56, 151–152] As shown in (3.114) and (3.115), the mass and inertia need to be adjusted by the square of the speed ratio. Based on this, the load torque required at the motor is

$$a_m = a_L \times R \tag{5.34}$$

$$I_m = I_L \times R^2 \tag{5.35}$$

$$T_m = I_m \times a_m \tag{5.36}$$

where $R = \dfrac{\text{Load Speed}}{\text{Motor Speed}}$

a_m = load acceleration reflected to the motor

a_L = load acceleration

I_m = load inertia reflected to the motor

I_L = load inertia

T_m = required motor torque

The same approach needs to be used on all components between the load and motor. If the engineer fails to account for all inertia in the system, the selected motor may not have adequate torque to perform as required. Motor torque to manage unbalance loads and acceleration loads should remain below the envelop defined by the peak stall torque/speed at peak power, peak torque at maximum speed/maximum speed as shown in Figure 5.7 or 5.8 for the defined duty cycle. [Hulst and Kamstra 2012, 11]

5.4.1.3 Motor Speed Requirements

Motor speed range needs to be considered to make sure both minimum and maximum actuator velocity requirements are satisfied. When motor speed is increased or decreased from load speed requirements to better match a motor's torque range, the engineer needs to make sure the motor is still operating within its maximum and minimum speed capability. The maximum motor speed required to achieve the stated maximum velocity is usually easy to determine, but any over speed conditions should also be considered to make sure the motor does not fail if a reasonable over speed condition is encountered. The minimum speed that the motor needs to operate smoothly at should also be specified to ensure stable operation. As available motor speed can depend on the torque required at that speed, the design engineer needs to make sure that the speed available aligns with the points and envelop as shown in Figure 5.7 or 5.8 for the defined duty cycle. Other requirements tend to be specific to motor type so they are addressed in the following sections.

5.4.1.4 Electric Motor Selection

Electric motors are popular prime movers for many applications. There are several types to choose from, allowing the engineer to choose a motor that matches the power requirements, control characteristics, cost constraints, and available power source(s).

Motor types include ac induction motors, ac induction with capacitor start motors, dc stepper motors, dc brushless motors, and dc brushed motors. Of these, the most common motors used in high-performance actuators are the dc stepper and brushless motor types because their speed and/or position can be precisely controlled with current technology. Successful application of electric motors requires careful matching of torque and speed requirements of the application to the torque and speed capability of the motor. When defining an electric motor for a drive application, the designer should try to cover the following items:

- Operating voltage.
- Current.
- Control system.
- For a stepper motor.
    - Step angle or number of steps.
    - Maximum and operating input pulse rate (pulses/second).
    - Maximum pull-in pulse rate.
- Peak stall torque.
- Peak torque at maximum speed.
- Continuous stall torque.
- Continuous rated torque.
- Maximum speed.
- Rated speed.
- Speed at peak power.
- Acceleration and deceleration time.
- Drive load (inertia and acceleration/deceleration or torque).
- Operating environment.
    - Temperature range (operating, storage).
    - Moisture (rain, humidity, condensation, . . .).
    - Contamination (dust, mud, cleaning compounds, . . .).
    - Shock and vibration.
- Allowable temperature rise.
- Motor shaft connection.
- Motor shaft radial loads (maximum, rated, speed).
- Motor shaft axial loads (maximum, rated, speed).
- Installation orientation.

- Size.
- Weight.
- Electrical connection.
- For a stepper motor, the following specifications are met:
  - step angle or number of steps
  - maximum and operating input pulse rate (pulses/second)
  - maximum pull-in pulse rate.

A general torque/speed graph for an electric motor is shown in Figure 5.7.

5.4.1.4.1 AC induction (synchronous) motors

AC induction motors use the induction of a magnetic field created in the rotor by the rotating field of the stator to create motion. The torque produced is a function of the power consumed and the speed is a function of alternating current supply frequency and the number of motor poles. Single-phase induction motors require some method of starting the motion. This can be accomplished by manually rotating the motor or using some sort of starting circuit. Capacitor start motors utilize another stator winding with an out of phase current created by a capacitor to generate a rotating magnetic field. As the number of poles is generally fixed, the speed of induction motors is varied by changing the ac supply frequency. [Nof 1985, 1284]

5.4.1.4.2 DC stepper motors

Stepper motor design is similar to an ac synchronous motor where an internal rotor with permanent magnets is controlled by external magnets that are switched electronically. Stepper motors do not have brushes eliminating maintenance associated with brushes. In a stepper motor, the rotor aligns with the magnetic field produced when each coil of the motor is energized. The motor "steps" from one position to the next as the windings are energized and de-energized in sequence. Step size can range from less than 1°–180°. By regulating how the coils are energized, motor direction, speed and position can be controlled so that they can be used a variable speed synchronous motor within the frequency range of the motor. They can generate large torques compared with the rotor moment of inertia and can generate large restoring torques if an external load tries to disturb a stable position. As stepper motors usually have a high power output to motor size ratio, motor temperatures can rise significantly under continuous use, so cooling or heat sinks should be considered for high-power and high-cycle applications. The torque-pulse rate curve for a typical stepper motor may look like the one shown in Figure 5.8. [Nippon Pulse America Inc. 2012, 1–4], [Ogata 1992, 217–224]

Figure 5.9 The dc stepper motor—the Kollmorgen PMX Series stepper motor (www.kollmorgen.com). Image courtesy of Kollmorgen.

5.4.1.4.3 DC brushless motors

DC brushless motors have rotating permanent magnets and a fixed armature, which eliminates the issues related to connecting current to a rotating armature. They have better efficiency, more torque per motor weight, more power, and less electromagnetic interference than brushed dc motors (Figures 5.10–5.12). Brushless motor reliability and life is better than brushed dc motors because wear and contamination issues associated with motor brushes are eliminated because they have permanent magnets that rotate and a fixed armature where current is connected to the motor without the need for moving parts. The fixed armature has no centrifugal forces on it and the windings are supported by the housing, which can also be cooled through active or passive conduction. These cooling options permit significant heat to be removed from the motor while allowing it to be completely sealed from the surrounding environment. As a result, brushless dc motors can handle high power levels without overheating, which can weaken motor magnets or shorten the life of other motor components. Some manufacturers have developed motors that can be integrated into the mechanical design to drive the load directly without additional mechanical transmission components. As the current is connected to a fixed armature, an electronic controller is needed to switch the phase to the armature windings to keep the rotor turning. Advances in electronic control systems have increased the reliability and reduced the cost of the controller, which performs timed power distribution instead of the brush/commentator system used in a brushed dc motor. Using a modern electronic control system to generate trapezoid or sine waveforms can reduce harmonic losses, compensate for winding induction, regulate power, and regulate monitor temperature. These characteristics make brushless dc motors a good choice for high-performance actuators. [Nof 1985, 35]

Figure 5.10 The dc brushless motor—Kollmorgen AKM Series Servomotor (www.kollmorgen.com). Image courtesy of Kollmorgen.

Figure 5.11 The direct drive dc brushless motor—the Kollmorgen Cartridge DDR motor (www.kollmorgen.com). Image courtesy of Kollmorgen.

Figure 5.12 The dc brushed motor—the Kollmorgen direct drive dc torque motor (www.kollmorgen.com). Image courtesy of Kollmorgen.

5.4.1.4.4 DC brushed permanent magnet motors

DC brush motor operation is similar to a dc brushless motor except that in brushed motors, the permanent magnets are fixed and the armature rotates. Brushes transfer the current from the fixed wiring to the rotating armature when they contact plates on the armature. As current passes through the coil on the armature, a magnetic field is generated causing the armature to be pushed away from the magnet with a like charge and drawn to a magnet with the opposite charge. As the magnetic field in armature aligns with the permanent magnets, the direction of the current through the armature is reversed when the brushes contact a different set of contact plates on the armature. This process continues as long as power is supplied to the motor. Such motors are inexpensive but can have lower reliability and increased maintenance requirements because of wear and contamination from the brushes. Also, heat from resistance to the electrical current passing through the coils in the armature cannot be readily conducted away. In general, the speed o of dc brushed motors is controlled by varying the applied voltage and torque is controlled by current. [Nof 1985, 35]

5.4.1.5 Hydraulic Motor Selection

Hydraulic motors offer the benefits of high power density and the ability to continuously control motor temperature by regulating the temperature of the hydraulic fluid. To obtain the required performance, the motor must be matched to pressure, flow, and thermal characteristics of the hydraulic power supply to which it is mated. Hydraulic motor torque is related to the pressure at the motor and motor displacement according to the following equation (Figure 5.13):

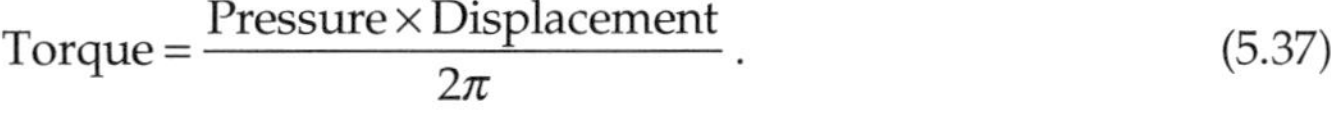

$$\text{Torque} = \frac{\text{Pressure} \times \text{Displacement}}{2\pi}. \tag{5.37}$$

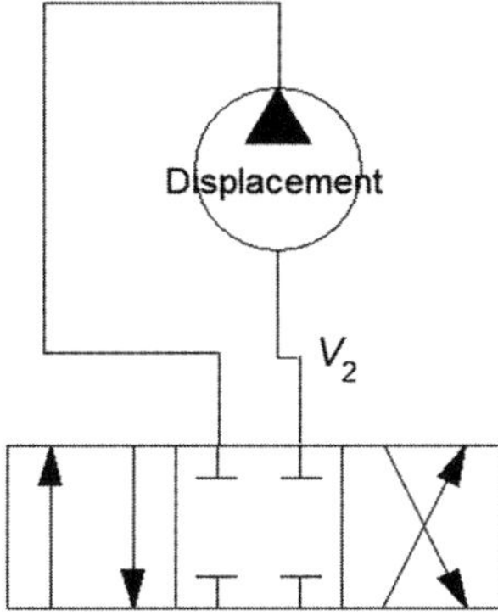

Figure 5.13 The hydraulic motor system.

Hydraulic motor speed is a function of motor displacement and flow rate as shown in the following equation:

$$\text{Speed}\left(\frac{\text{revolutions}}{\text{time unit}}\right)=\frac{\text{Rate}}{\text{Displacement}}$$
$$\text{Speed}\left(\frac{\text{radians}}{\text{time unit}}\right)=\frac{2\pi\times\text{Flow Rate}}{\text{Displacement}}. \tag{5.38}$$

As power is a function of torque and speed, hydraulic motor power is represented by

$$\text{Power}=\text{Torque}\times\text{Speed}\left(\frac{\text{radian}}{\text{time unit}}\right)$$
$$\text{Power}=\text{Pressure}\times\text{Flow Rate}. \tag{5.39}$$

Hydraulic motor power in metric units is given by

$$\text{Power (kW)}=\text{MPa}\times\frac{1E^{6}\ \text{Pa}}{\text{MPa}}\times\frac{\frac{\text{N}}{\text{mm}^{2}}}{\text{Pa}}\times\frac{\text{L}}{\text{min}}\times\frac{\text{mm}^{3}}{\text{L}}\times\frac{\text{m}}{1E^{3}\ \text{mm}}\times\frac{\text{min}}{60\ \text{s}}$$
$$\times\frac{\text{J}}{\text{N}\times\text{m}}\times\frac{\text{W}\times\text{s}}{\text{J}}\times\frac{\text{kW}}{1E^{3}\ \text{W}} \tag{5.40}$$
$$\text{Power (kW)}=\frac{\text{MPa}\times\frac{\text{L}}{\text{min}}}{60}.$$

In English units, it is given by

$$\text{Power (HP)}=\frac{\left(\frac{\text{lb}}{\text{in}^{2}}\right)\times\left(\frac{\text{gal}}{\text{min}}\right)\times\frac{231\ \text{in}^{3}}{\text{gal}}\times\frac{\text{f}}{12\ \text{in}}}{33,000\frac{\text{f}\times\text{lb}}{12\times33,000}}$$
$$\text{Power (HP)}=\frac{231\times\text{psi}\times\text{gpm}}{12\times33,000} \tag{5.41}$$
$$\text{Power (HP)}=\frac{\text{psi}\times\text{gpm}}{1714}.$$

[Sperry Vickers 1970, 6-10 to 6-11], [The Lee Company 2000, 18]

An important characteristic for controlling hydraulic motors in high-performance actuators is motor stiffness as it contributes to overall system stiffness. Stiffness of a hydraulic motor is directly related to displacement and hydraulic fluid bulk modulus. Of these two factors, the design engineer has the greatest control over motor displacement. Hydraulic motor stiffness can be calculated as follows:

- Displacement = motor displacement per revolution
- β = fluid bulk modulus
- P = pressure
- V = volume
- θ = motor rotation
- V_1 = total trapped hydraulic fluid volume on motor side 1

- V_2 = total trapped hydraulic fluid volume on motor side 2
- T = torque
- J = total moment of inertia of system

Recalling from $\beta = \frac{V_o \times (P_o - P_1)}{V_o - V_1}$ and rearranging

$$\Delta V = \frac{V_o \times \Delta P}{\beta}$$
$$\Delta\theta = \frac{\Delta V}{\text{Displacement}}$$
$$\Delta\theta_{\text{radian}} = 2\pi\left(\frac{\Delta V}{\text{Displacement}}\right) \tag{5.42}$$
$$\Delta\theta_{\text{radian}} = \left(\frac{V_o \times \Delta P}{\beta}\right)\left(\frac{2\pi}{\text{Displacement}}\right).$$

From (5.37)

$$T = \frac{P \times \text{Displacement}}{2\pi}$$
$$\Delta T = \frac{\Delta P \times \text{Displacement}}{2\pi}$$
$$\Delta P = \frac{2\pi \times \Delta V}{\text{Displacement}} \tag{5.43}$$
$$\Delta\theta_{\text{radian}} = \left(\frac{2\pi}{\text{Displacement}}\right)\left(\frac{V_o}{\beta}\right)\left(\frac{2\pi \times \Delta T}{\text{Displacement}}\right)$$
$$\frac{\Delta T}{\Delta\theta_{\text{radian}}} = \frac{\beta \times \text{Displacement}^2}{4\pi^2 V_1}$$

which is hydraulic motor stiffness (spring rate).

For a valve controlling fluid in and out of the motor, the fluid acts like two springs in parallel. The final spring rate (total hydraulic motor spring rate) for this configuration is

$$k = \frac{\Delta T}{\Delta\theta_{\text{radian}}} = \left[\frac{\beta \times \text{Displacement}^2}{4\pi^2 V_1}\right] + \left[\frac{\beta \times \text{Displacement}^2}{4\pi^2 V_2}\right]. \tag{5.44}$$

The hydraulic motor natural frequency for this motor is

$$\omega_o = \sqrt{\frac{k}{J}}. \tag{5.45}$$

[Rexroth Worldwide Hydraulics 1984, 8]

There are several types of hydraulic motors that can be used in high-performance drives. They can be grouped into two major categories: fixed displacement and variable displacement. As one can see from (5.44), variable displacement motors do not work well in high-performance applications because their stiffness and resulting natural frequency decreases as the displacement decreases. Likewise, one can see that while high operating pressure can minimize the displacement needed to develop the required motor torque, but low motor displacement may make the system natural frequency low. If the displacement needs to increase just to satisfy motor stiffness requirements, then energy is wasted, usually in the form of heat. Selection of the motor depends on many variables such as required displacement, shaft loading, operating pressure, speed, and company or customer preferences or standard equipment. In addition, the motor has to be compatible with the hydraulic fluid available and fluid characteristics, such as viscosity, cleanliness, lubricity, and flammability, need to be considered when selecting the motor. Within the fixed and variable displacement families motor types include gear, vane, gerotor, axial piston, and radial piston. A short description of sample motor types follows.

5.4.1.5.1 Gear motors

Gear motors consist of a housing with two gears that form a low leak mesh between each other and have a small clearance between the tip of the gear teeth and the mating housing cavity (Figure 5.14). High-pressure fluid enters a chamber on one side of the gear mesh, causes the gears to rotate, and travels around the gears until it reaches the low pressure fluid return chamber on the other side of the gear mesh. Gear motors are considered to be high-reliability devices because they are tolerant to contamination and their failure mode is generally due to high leakage caused by component wear. They may not operate at the speed or pressure levels that a piston motor does, but a high-quality gear motor can offer smooth, reliable operation for several years.

Figure 5.14 The gear motor—the Parker PGP/PGM 500 Series gear motor (www.parker.com). Image courtesy of Parker Hannifin Gear Pump Division.

5.4.1.5.2 Vane motors

Vane motors have a rotor with vanes that is mounted in an eccentric bore in the motor housing (Figure 5.15). The vanes slide in and out of slots in the rotor as it turns. Pressurized fluid enters a chamber in the motor and acts on the vanes causing the rotor to turn. As the rotor turns, the pressurized fluid moves to the low-pressure chamber where it is returned to the hydraulic tank. The vanes are pushed into the slot at the transition between the lower pressure chamber and the high-pressure chamber to ensure sealing and create the unbalance force that causes the rotor to turn. Vane motors offer good efficiencies and life at reasonable cost. They also are available in a large range of displacements and with low-speed/high-torque characteristics.

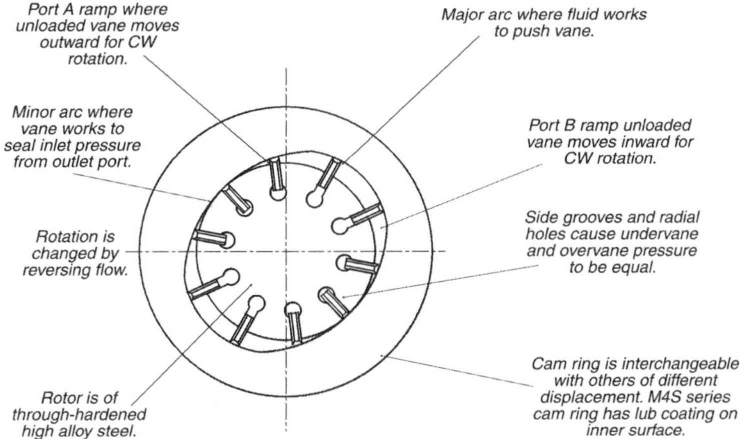

Figure 5.15 The vane motor—the Parker Series M3B - M4* vane motor (www.parker.com). Images courtesy of Parker Hannifin VPDE, Denison Vane Motors.

5.4.1.5.3 Gerotor/Geroler motors

Gerotor and geroler motors consist of a housing with two sets of gears whose centers are offset from each other. Both gerotor and geroler have the same basic design and operate on the same principle. Geroler motors have rollers on the ends of the outer gear lobes that reduce friction and improve motor efficiency. The outer gear has internal gear teeth (N) with one more tooth than the outer gear teeth of the inner gear ($N - 1$). Tooth form is defined so that all teeth on the inner gear are always in contact with the outer gear. Each engagement between a tooth of the inner gear and the outer gear represents one pressure to return fluid cycle. The inner gear also has an internal gear that mates with the output shaft, causing it to rotate as the inner gear moves around the internal gear teeth of the outer gear. Gerotor motors provide economical speed and torque performance in the lower to middle operating ranges because they lose efficiency as flow and pressure increase (Figure 5.16).

Figure 5.16 The Gerotor motor—the Parker TC Series hydraulic motor (www.parker.com). Image courtesy of Parker Hannifin Hydraulic Pump/Motor Division.

5.4.1.5.4 Axial and bent axis piston motors

Axial and bent axis piston motors utilize a series of pistons arranged in a circular pattern around the axis of rotation and attached to a swash plate (Figure 5.17). As high-pressure fluid enters the bore, it pushes the piston causing the cylinder block with pistons to rotate. A valve plate on the open end of the piston bore contains the inlet and outlet. The inlet coincides with the piston positions that are being pushed out of the bore and the outlet coincides with piston positions being pushed into the bore. Axial and bent axis piston motors are considered some of the highest quality, highest efficiency motors, and can have a very large speed range, operating from below 50 rpm to several thousand revolutions per minute. However, as with all hydraulic motors, efficiency, torque, and speed characteristics are dependent on the design approach used.

Figure 5.17 The bent axis piston motor—the Parker F1 bent axis piston motor (www.parker.com). Image courtesy of Parker Hannifin Hydraulic Pump and Motor Division Europe.

5.4.1.5.5 Radial piston motors

Radial piston motors utilize pistons arranged radially around the drive shaft that reciprocate perpendicular to the axis of rotation (Figure 5.18). The pistons can be attached to a crankshaft, which turns as pressurized fluid flows into the bore of an extended piston. When a piston reaches the bottom of its stroke, the bore is connected to the hydraulic return line so that the fluid can be returned to tank as the piston moves back to the extended position. Radial piston motors have high reliability and high efficiency and are capable of producing very high starting torques. As such, they can produce very high power but are not well suited to high-speed operation. [Sperry Vickers 1970, 6-6 to 6-28], [John Deere 1979, 5-2 to 5-10]

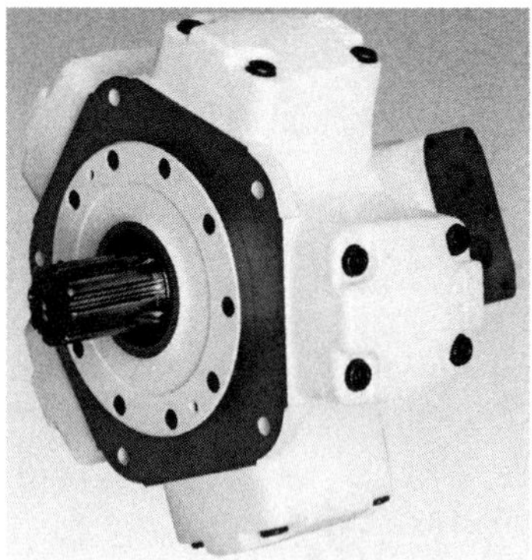

Figure 5.18 The radial piston motor—the Parker Calzoni MR Series radial piston motor (www.parker.com). Image courtesy of Parker Hannifin Hydraulic Pump Division.

5.4.1.6 Pneumatic Motors

Pneumatic motors can be a popular option for some systems where an adequate air supply is readily available or where there are safety concerns with electrical or hydraulic power (Figure 5.19). Another advantage to pneumatic motors is that no fluid return line is required as the low-pressure fluid can usually be discharged to the immediate environment with only sound and pressure conditioning. However, one needs to be aware of the low stiffness and resulting low natural frequency that can accompany pneumatic drives. This is due to the low bulk modulus of gases. Hydraulic fluids can have a bulk modulus in the range of 1400 to 2000 MPa (200,000 to 300,000 psi), while air has a bulk modulus of 28 MPa (4000 psi) or 1.5% to 2% of a hydraulic fluid. This can make designing stable high-performance pneumatic systems a challenge, but the basic design principles still apply and a stable system can be developed. Calculations and equations for pneumatic motor sizing are different from those for hydraulic motors. The designer needs to utilize the appropriate physics for gases. Traditionally, pneumatic motors require some sort of lubricant to be introduced into the fluid stream to keep wear, corrosion, heat, and seizing of motor components to a minimum. Nowadays, many air motors run without air lubrication. This is valid for small-to-medium-sized vane motors and turbines. The operating principles for pneumatic motors are similar to hydraulic motors and motor types are similar. Some design features are different to account for operating fluid differences. Air motor types include turbine, vane, axial piston, and radial piston, and gear motors. Air turbine motors such as used in dentist drills can achieve very high speeds but develop little torque. Vane motors are relatively simple motors and can achieve high speeds and reasonable torques. Axial and radial piston motors can generate relatively high torques but operate at lower speeds due to reciprocating component mass. [Gast Manufacturing Corporation 1986, 8–13]

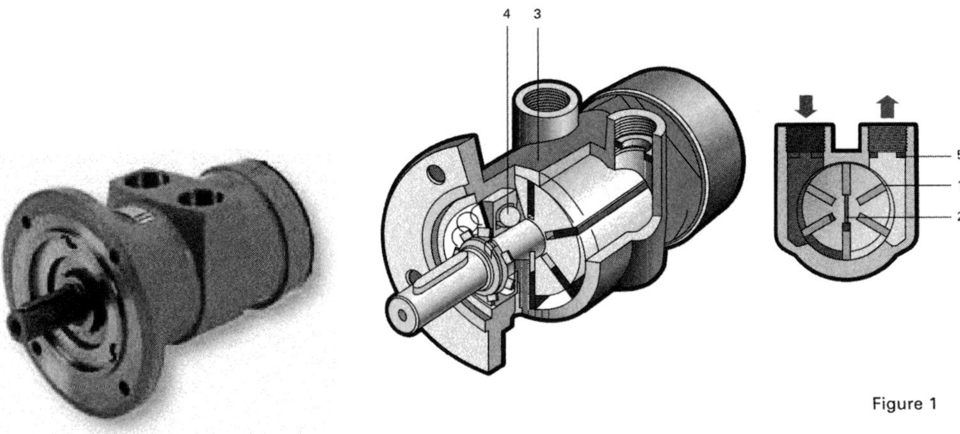

Figure 5.19 The pneumatic motor—the Atlas Copco LZL Series pneumatic vane motor (www.atlascopco.com).

5.4.2 References

Shortley, G., and Williams, D. 1971. Elements of Physics, Fifth Edition. Prentice-Hall, Inc: Englewood Cliffs, NJ.

John Deere Service Publications. 1979. Hydraulics, Fundamentals of Service. Deere & Company: Moline, IL.

Sperry Vickers. 1970. Industrial Hydraulics Manual. Sperry Rand Corporation: Troy, MI.

The Lee Company Technical Center. 2000. Technical Hydraulic Handbook. The Lee Company: Westbrook, CT.

Rexroth Worldwide Hydraulics. 1984. Components for Hydraulic Proportional and Servo Systems, Electronics and Accessories. The Rexroth Corporation: Bethlehem, PA.

Nippon Pulse America, Inc. 2012. Basics of Steppper Motors. Nippon Pulse America, Inc: Radford, VA.

Hulst, S., and Kamstra, L. 2012. Gearmotors: Achieving the Perfect Motor & Gearbox Match. Groschopp, Inc: Sioux City, IA.

Gast Manufacturing Corporation. 1986. Air Motors Handbook. Gast Manufacturing Corporation: Benton Harbor, MI.

Nof, S. Y. 1985. Handbook of Industrial Robotics. John Wiley & Sons, Inc: New York.

Ogata, K. 1992. System Dynamics. Prentice Hall: Englewood Cliffs, NJ.

5.5 Control Element Design and Selection

Design principles associated with controlling high-performance drives is similar for all types of power systems, whether electrical, hydraulic, or pneumatic. The mechanical control elements must work seamlessly with any associated electronic or software controls to create a robust, stable and safe system that is economically feasible. This book addresses the fundamentals of actuator design from a mechanical perspective; the reader should consult other sources for additional information on control element design and selection.

5.6 Linear Actuator Design

When high-performance linear motion is required, there are five primary types of actuators that can be used: cylinders, motor and rack, screws, electric solenoids, and electric linear motors. Within each of these types, the engineer can select from several basic designs based on power source, performance requirements, design life, and cost.

Structural requirements for a linear actuators include making sure that they can carry the load, survive for their intended life, and have the stiffness needed for controllability. With today's computer-aided design tools, one can perform sophisticated structural analysis including fatigue analysis to determine if they will survive for their intended life. The key to performing good analyses and getting applicable results is understanding the applied load(s). A good starting point for design is to determine the elastic buckling load for the column member. The basic equation for a column with pinned ends as shown in Figure 5.20 is developed below.

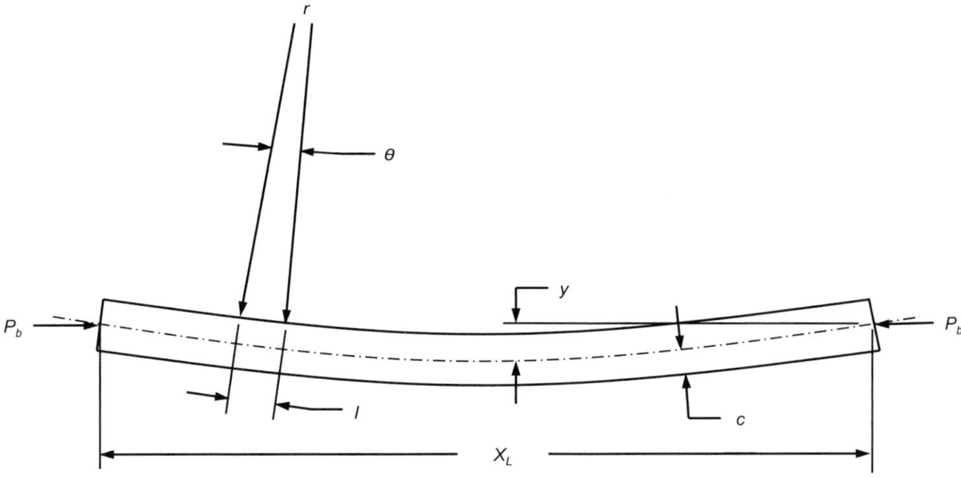

Figure 5.20 The basic elastic beam with pinned ends.

For the basic elastic column shown in Figure 5.20, at any given point on the curve θ

$$\text{Slope} = \tan\theta = \frac{dy}{dx}. \tag{5.46}$$

If $\frac{dy}{dx}$ is small, then

$$\tan\theta = \frac{dy}{dx} \approx \theta \tag{5.47}$$

$$\text{Curvature} = \frac{1}{r} = \frac{d\theta}{dx}. \tag{5.48}$$

For

$$\text{Stress} = \sigma = \frac{Mc}{I} \tag{5.49}$$

where $M = Pb_b \times y$ and c = distance from the neutral axis and I = area moment of inertia for the beam.

For

$$\text{Strain} = \varepsilon = \frac{\Delta l}{l} \tag{5.50}$$

where Δl = change in length and l = length.

Knowing

$$E = \frac{\sigma}{\epsilon} \tag{5.51}$$

where E = elastic modulus.

Combining (5.49) and (5.51), then rearranging

$$\epsilon = \frac{Mc}{EI}. \tag{5.52}$$

For

$$\tan\theta = \frac{l}{r}. \tag{5.53}$$

For small deflections at the neutral axis

$$\theta = \frac{l}{r}. \tag{5.54}$$

At the other fivers

$$\theta = \frac{l_1}{r+c} \tag{5.55}$$

$$\Delta l = l_1 - l \tag{5.56}$$

$$\Delta l = \theta(r+c) - \theta r \tag{5.57}$$

$$\Delta l = c\theta. \tag{5.58}$$

Combining (5.50), (5.52), and (5.58)

$$\frac{c\theta}{l} = \frac{Mc}{EI} \tag{5.59}$$

$$\frac{\theta}{l} = \frac{M}{EI}. \tag{5.60}$$

Combining (5.54) and (5.60)

$$\frac{\theta}{r\theta} = \frac{M}{EI} \tag{5.61}$$

$$\frac{1}{r} = \frac{M}{EI}. \tag{5.62}$$

Combining (5.48) and (5.62), we get

$$\frac{d\theta}{dx} = \frac{M}{EI}. \tag{5.63}$$

From (5.47), we get

$$\frac{d\theta}{dx} = \frac{d^2y}{dx^2} \,. \tag{5.64}$$

Combining (5.63) and (5.64) gives the general elastic deflection equation and the elastic bulking load can be determined from the general elastic deflection equation:

$$EI\frac{d^2y}{dx^2} = M \,. \tag{5.65}$$

This can be used to determine the elastic buckling load as follows.

If there is a deflection in the beam shown in Figure 5.20, the moment at any point of the beam is

$$M = P_b y \tag{5.66}$$

where y = deflection from the center and P_b = compressive load applied to the beam

So

$$EI\frac{d^2y}{dx^2} = P_b y \tag{5.67}$$

or

$$EI\frac{d^2y}{dx^2} - P_b y = 0 \,. \tag{5.68}$$

The solution to this equation is of the form

$$C_1 \sin kx + C_2 \cos kx = 0 \,. \tag{5.69}$$

For no deflection at the beam ends

$$C_1(0) + C_2(0) = 0 \,.$$

So

$$y = C_1 \sin kx \tag{5.70}$$

$$\frac{dy}{dx} = kC_1 \cos kx \tag{5.71}$$

$$\frac{d^2y}{dx^2} = -k^2 C_1 \sin kx \,. \tag{5.72}$$

Combining (5.68) and (5.72), we get

$$EI\left[-k^2 C_1 \sin kx\right] - P_b\left[C_1 \sin kx\right] = 0 \tag{5.73}$$

$$\left[C_1 \sin kx\right]\left[-EIk^2 - P_b\right] = 0 \tag{5.74}$$

which is true when

$$\left[C_1 \sin kx\right] = 0 \text{ or} \left[-EIk^2 - P_b\right] = 0$$
$$k = \pm\sqrt{\frac{P_b}{EI}}. \tag{5.75}$$

If $x = 1$, $y = 0$ and the first condition is

$$y = 0 = C_1 \sin\sqrt{\frac{P_b}{EI}}L. \tag{5.76}$$

As $C_1 \neq 0$

$$\sqrt{\frac{P_b}{EI}}L = n\pi \tag{5.77}$$

$$P_b = \frac{(n\pi)^2 EI}{L^2} \tag{5.78}$$

where the smallest value for P_b is for $n = 1$

$$P_b = \frac{\pi^2 EI}{L^2}. \tag{5.79}$$

Equation (5.79) represents the elastic buckling load for a beam with pinned ends. For other end conditions, one can look at how the effective length of the buckling mode compares with the length (L) for the simply supported beam. A generalized form of $P_b = \frac{\pi^2 EI}{L^2}$ is beam buckling load

$$P_b = \frac{\pi^2 EI}{L_e^2} \tag{5.80}$$

where L_e is the effective length (equivalent to half-sine curve cycle for the configuration) [Byars and Snyder 1975, 319–329] (Figure 5.21)

Column Equivalent Length (L_e)		
End Condition	L_e	
Pinned Ends	$L_e = L$	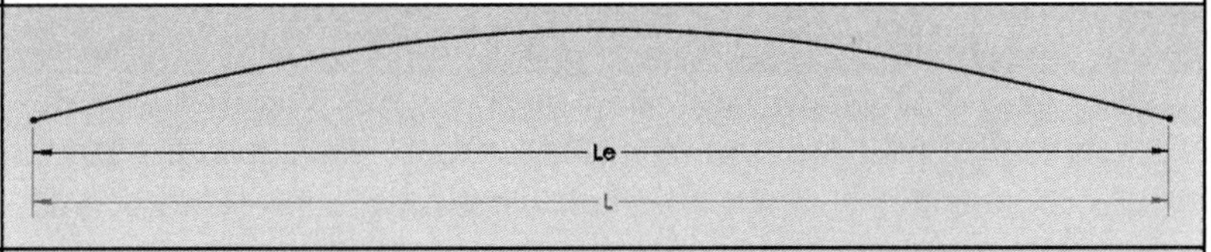
Fixed Ends	$L_e = .5L$	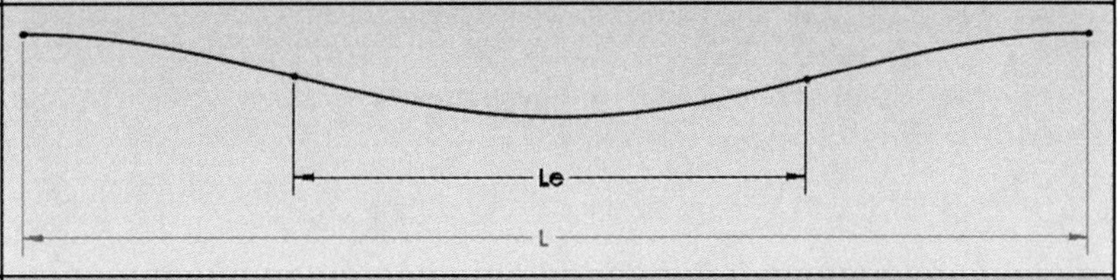
One End Fixed/ One End Free	$L_e = 2L$	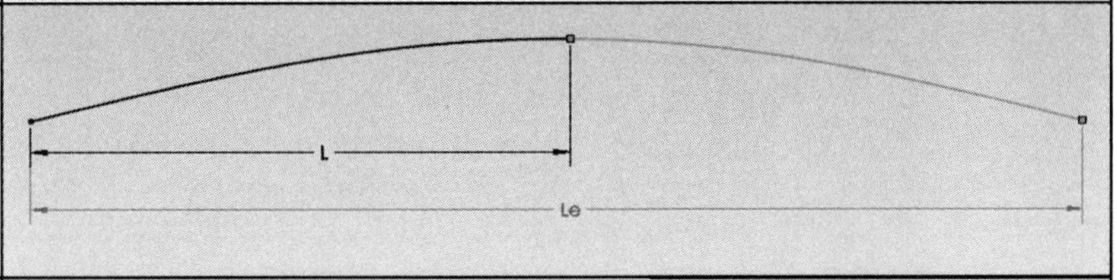
One End Fixed/ One End Pinned	$L_e = .7L$	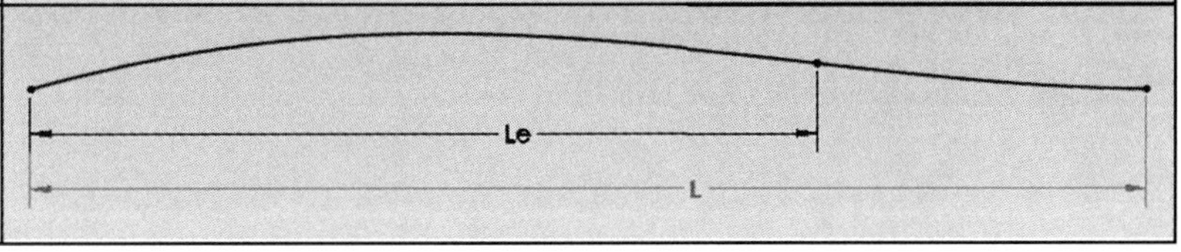

Figure 5.21 Beam equivalent lengths.

5.6.1 Cylinders

Cylinders are generally composed of an outer body with an inner piston connected to a rod that is moved by a pressurized fluid. [Sperry Vickers 1970, 6-1 to 6-6], [John Deere 1979, 4-1 to 4-6] The cylinder general arrangement and components is shown in Figure 5.22.

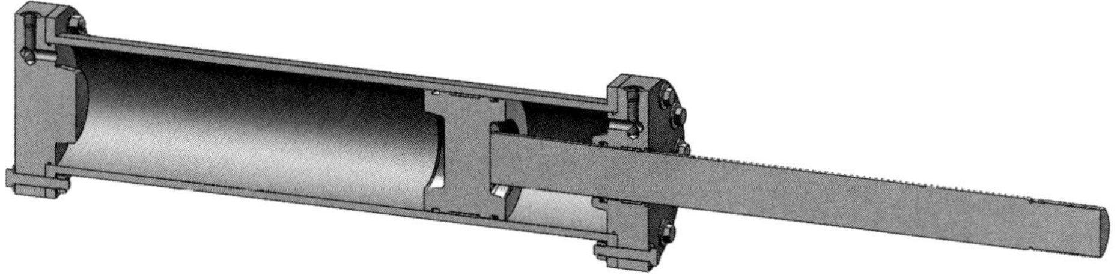

Figure 5.22 The cylinder.

The most common fluids used to operate cylinders are hydraulic fluid and air. Selection of which fluid to use is primarily set by the application and the fluid available. In general, hydraulic fluids are used when forces are high and precise control is needed. Common hydraulic applications are mobile equipment and large equipment requiring large forces. Air is common in industrial settings where air is readily available, and force and/or control requirements are low. General equations defining cylinder performance with an incompressible fluid are presented below. If the fluid must be considered compressible, then the rules for compressible fluids must be applied.

The fluid volume required V can be represented as

$$V_p = A_p \times S$$
$$V_r = A_r \times S$$

where V_p = piston side volume

V_r = rod side volume

A_p = piston side area

A_r = rod side area

S = cylinder stroke

The cylinder velocity v can be represented as

$$v_p = \frac{Q_p}{A_p}$$

$$v_r = \frac{Q_r}{A_r}$$

and

$$v_p = v_r$$

where v_p = piston side velocity

v_r = rod side velocity

Q_p = piston side flow rate

Q_r = rod side flow rate

A_p = piston side area

A_r = rod side area

S = cylinder stroke

The cylinder force F is given by

$$F = A_p P_p - A_r P_r - \mathcal{F}_p - \mathcal{F}_r$$

where F is positive when the rod is extending

A_p = piston side area

P_p = piston side pressure

A_r = rod side area

P_r = rod side pressure

$\mathcal{F}_p$ = piston seal friction

$\mathcal{F}_r$ = rod seal friction

As seen in these equations, one has to be sure to account for all losses in the system to accurately predict actual cylinder performance. Seal friction and return side pressure are common items to ignore during initial sizing calculations, but this omission can lead to an undersized design and actual performance that does not meet expectations. Other items to consider include supply pressure drop as velocity increases and return pressure increase as velocity increases, along with the difference between static and dynamic seal friction values. The latter can be a concern when trying to precisely control a cylinder and the control system has to deal with the "stiction" that can appear when the actuator needs to move a small amount from a zero velocity condition or is approaching a desired position at low velocity and friction causes the cylinder to stop before it reaches the desired position.

The following sections address some design considerations specific to hydraulic and pneumatic cylinders.

5.6.1.1 Hydraulic Cylinders

Hydraulic cylinders are one of the most common actuators. They are inexpensive, reliable, and can exert high forces. In addition, they can provide high acceleration, which also makes them a favorite when high-performance is required. When mated to a hydraulic system with accumulators, the high-peak power levels can be averaged over the entire motion cycle to reduce the size of the power source. However, designing a high-performance hydraulic cylinder is not a trivial task. In addition to the structural requirements of the cylinder, the engineer must consider friction, load control, failure modes, and select the fluid.

The best mounting configuration for a cylinder is when both ends have pinned joints. Pinned joints ensure that no moment is transferred to the cylinder, and therefore, it is in either tension or compression. This minimizes side loads, and therefore, friction and wear on the rod and piston. If one or both ends of the cylinder need to be rigidly mounted, the design needs to incorporate bearings at both the piston and rod seal to carry any side load that may be induced from either a direct lateral load or moment transferred through the cylinder. Bearing material and size selection needs to be based on fluid compatibility, operating velocity, and load. Bearing materials are often rated by

PV, where *P* is the applied pressure and *V* is the velocity. Hence, a cylinder operating at high velocities cannot tolerate much side load, and conversely, a cylinder operating with a high side load cannot operate at high velocity. In addition to the tension, compression, and moment loads that the cylinder must withstand, the engineer must also consider the fluid pressure loads.

For high-performance cylinders, intermittent pressure loads can be significant. They can be higher than system operating pressure when decelerating a load and even greater than the system relief valve setting for brief periods when load inertia causes system pressure to rise faster than when the relief valve can open. If designing such a system, utilizing a small direct acting relief valve in close proximity to the load may offer the fastest response time. However, the system needs to be tested to determine the peak pressure reached due to either a system failure or unexpected load. Circumferential stress in the cylinder should be checked to make sure that it is below the material yield point by an appropriate factor of safety. Deflection of the cylinder also needs to be checked under pressure to make sure that the clearance between the piston and cylinder bore does not become too large. This gap can be governed by two factors, depending on the design approach. Some designs may not have a seal between the piston and cylinder bore because the friction and stiction cannot be tolerated to achieve the level of control required. In this case, any cylinder case deflection will increase leakage past the cylinder and make control more difficult, waste energy and create heat. In the case where there is a seal between the piston and cylinder bore, excessive case deflection can cause the seal to extrude out of its gland and then be caught between the piston and case when pressure decreases and the cylinder bore contracts. This increases the force required to move the piston. It also reduces seal life when parts of the trapped seal are torn off when the piston begins to move. The same issues can happen to the rod seal, but it usually has more material around it so that deflection is easier to manage.

The designer also needs to consider seal installation when designing the piston and rod end cap. There are two types of seal gland, as shown in Figures 5.23 and 5.24, open and closed. Most high-performance, high-pressure seals are made from stiff materials that can tolerate pressure and velocities yet minimize friction. These materials are generally not flexible and therefore want an open gland. Open glands require more space and components so space for them needs to be allocated during the concept phase or else there may not be adequate room to fit them in during detailed design. When the sealing system is smaller and more flexible, relative to the bore and rod size, closed glands may be utilized. These are much simpler and therefore preferred if seal installation will tolerate them. One other item to consider when designing for seal installation is the appropriate lead-in angle for the seal. These are required on the piston leading to the piston seal gland, piston bore, rod, and rod end cap, leading to the rod seal gland. These can be accomplished by either providing the required lead-in on the piston, piston bore, rod, and rod end cap or utilizing an installation tool that provides the proper lead-in.

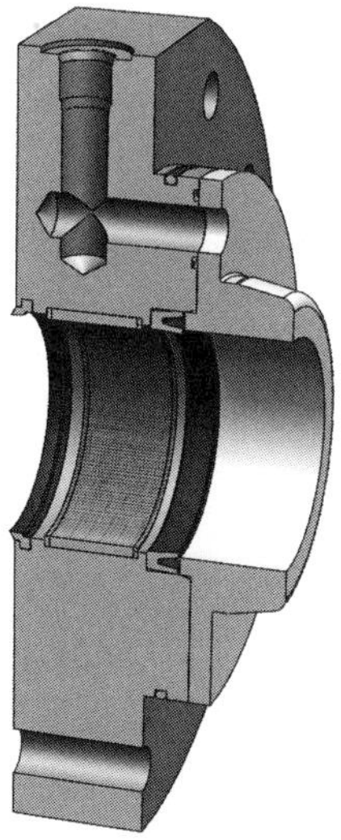

Figure 5.23 The open seal gland.

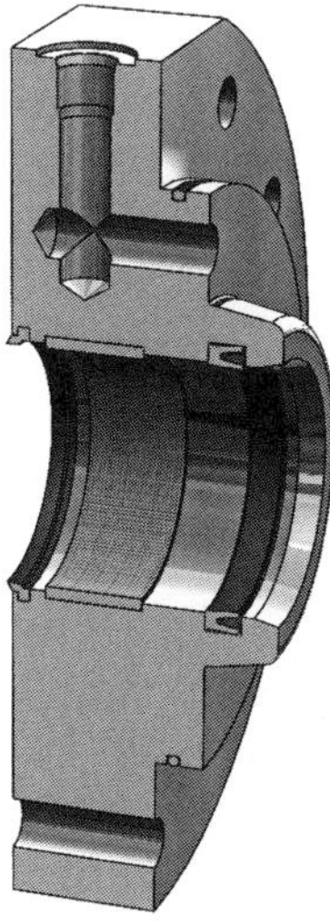

Figure 5.24 The closed seal gland.

Specific seal selection should be done based on previous experience with seals in similar applications or consulting with seal suppliers to identify potential seal designs. If the opportunity is available, the most promising seal designs should be tested in the application to determine how well they meet performance, life, and cost objectives. Seal designs available may include single, dual, multipart, vented, and nonvented. In some cases, multiple seals or multipart seals may offer significant performance and life advantages but carry disadvantages in cost and potential for incorrect assembly. All these factors along with any actual test data should be weighed and evaluated during final seal selection.

The final item to be considered as part of seal design is seal lubrication requirements. Seal lubrication requirements depend on seal material, operating fluid, operating loads, and operating velocities. The best source of design information is the seal supplier. In some cases, one can get by with the as-finished rod or bore surface; if the surface provides a low enough coefficient of friction, the operating fluid has adequate lubricating properties, and/or the surface can trap enough fluid to lubricate one operating cycle. If the operating fluid does not supply adequate lubrication, a special seal lubricant can be injected at the seal interface to minimize wear and provide a sealing film. This can be accomplished with a separate tank of seal fluid that is pressurized by the operating system as shown in Figure 5.25. If the seal mating surface does not trap enough lubricant, the surface plating can be changed to a material/process that provides more cavities for fluid to be retained or the surface can be modified, usually by grinding, to provide places for fluid to be trapped. One of the best surfaces is the cross-hatched pattern as shown in Figure 5.26.

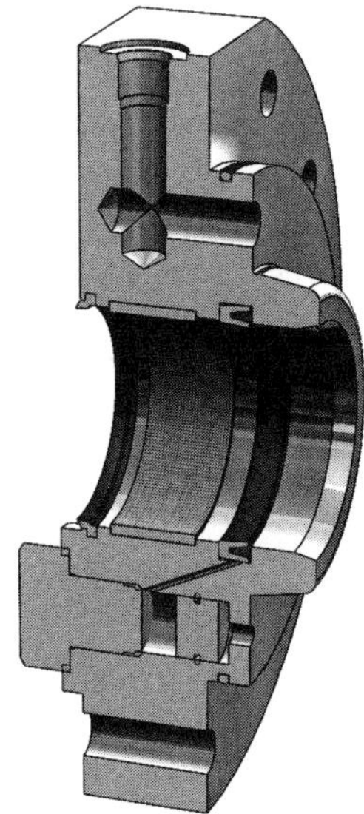

Figure 5.25 Lubricated seal.

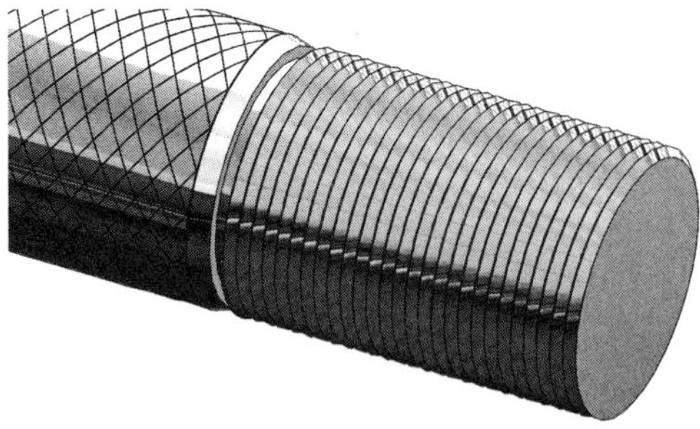

Figure 5.26 Cross-hatched seal surface.

For high-performance hydraulic cylinders, the engineer needs to consider how the cylinder will behave near the end of its stroke, both at the extended and retracted positions. In some cases, the cylinder is controlled externally by a control valve that operates it precisely over the entire stroke. In other cases, the cylinder just needs to operate with controlled motion at each end of its stroke. One advantage of hydraulic cylinders is the ability to add cushions at the end of the stroke that can dissipate a significant amount of energy. These cushions can be utilized as simple, robust deceleration controls for normal or fail-safe operation. For systems that have a control valve for normal operation, the cushions can be designed as a fail-safe feature if there is a failure in the primary control system. For systems that need only precise acceleration and/or deceleration control at the end of their stroke, cushions can provide a reliable, cost effective method to accomplish the control required. Some examples of cushion design are shown in Figures 5.27–5.29.

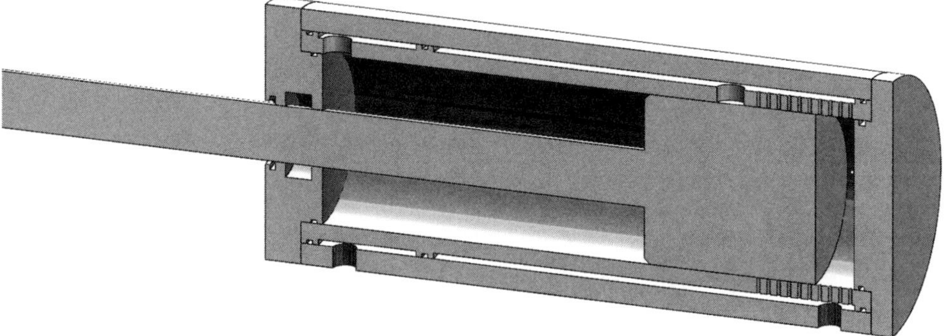

Figure 5.27 Piston cushion with holes.

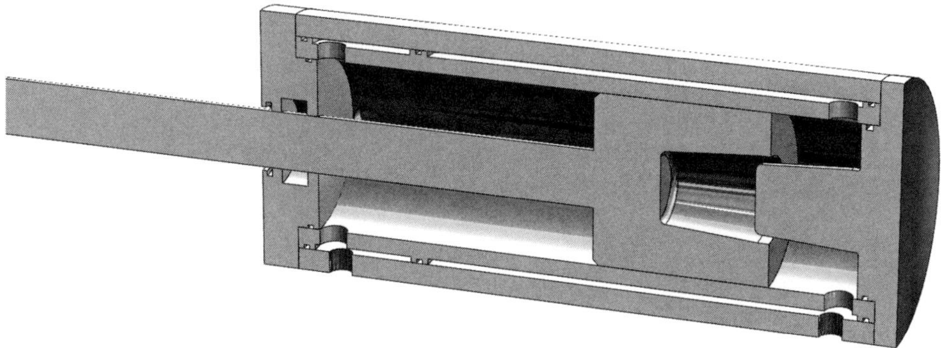

Figure 5.28 Piston cushion with tapered rod.

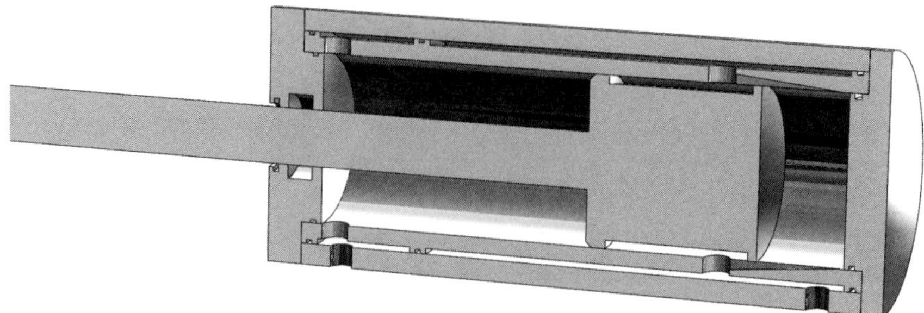

Figure 5.29 Piston cushion with grooves.

The basic design approach for cushions is to limit the flow into or out of the cylinder as a function of position so that the fluid pressure applied to the piston results in the desired acceleration or deceleration. The generalized equation for piston deceleration [Shortley and Williams 1971, 56] is

$$a = \frac{F}{m} \tag{5.81}$$

where a = deceleration rate, F = force generated, and $F = P \times A$

where A = piston area

for the piston side: $A_p = \frac{\pi \times d^2}{4}$, where d = piston diameter

for the rod side: $A_r = \frac{\pi \times (d^2 - d_r^2)}{4}$, where d = piston diameter and d_r = rod diameter

p = pressure applied to the piston and is a function of piston velocity

From Bernoulli's equation [Streeter and Wylie 1975, 134–140]: $C = \frac{v^2}{2} + gh + \frac{p}{\rho}$, we can derive the pressure drop as a function of flow rate, orifice size, and flow coefficient. If the head pressure is the same on both sides of the orifice, the upstream side of the orifice is governed by $\frac{P}{\rho}$ because the velocity component is small and the downstream side of the orifice is governed by $\frac{v^2}{2}$ because the pressure component is small and we get

$$\frac{P}{\rho} = \frac{v^2}{2}. \tag{5.82}$$

Knowing $v = Q/a_o$

$$\frac{P}{\rho} = \frac{(Q/a_o)^2}{2} \tag{5.83}$$

$$P = \frac{\rho}{2}\left(\frac{Q}{a_o}\right)^2. \tag{5.84}$$

Equation (5.84) is the theoretical flow through the area of the orifice hole, but depending on entrance and exit effects, the actual flow area can be smaller. Adding an orifice coefficient C to (5.84) allows it reflect actual flow conditions for a given hole configuration:

$$P = C\frac{\rho}{2}\left(\frac{Q}{a_o}\right)^2. \tag{5.85}$$

We can utilize an orifice coefficient C_o that combines the half term and the other orifice characteristics to simplify (5.84) into a final equation representing pressure drop across an orifice

$$P = C_o\rho\left(\frac{Q}{a_o}\right)^2 \tag{5.86}$$

where C_o = orifice coefficient

ρ = fluid density

Q = fluid flow rate

a_o = orifice area ($a_o = \pi r^2$)

For a simple incompressible fluid case where the fluid can be considered to flow through a single orifice

$$P = \left[\frac{Q}{C_{om}a}\right]^2 + p_r \tag{5.86a}$$

where p_r = return line pressure

a = orifice area

C_{om} = modified orifice coefficient $\left(\sqrt{\frac{1}{C_o\rho}}\right)$

Q = flow rate from the piston ($Q = V_p \times A$), where V_p = piston velocity and A = piston area as defined above.

For "n" orifices in parallel, an equivalent orifice area can be represented by:

$$a_{ep} = a_1 + a_2 + \dots + a_n \tag{5.86b}$$

For "n" orifices in series, an equivalent orifice area can be represented by:

$$\frac{1}{(a_{es})^2} = \frac{1}{(a_1)^2} + \frac{1}{(a_2)^2} + \dots + \frac{1}{(a_n)^2} \tag{5.86c}$$

or:

$$a_{es} = \sqrt{\frac{1}{\frac{1}{(a_1)^2} + \frac{1}{(a_2)^2} + \dots + \frac{1}{(a_n)^2}}} \tag{5.86d}$$

To achieve a smooth deceleration rate, as discussed in section 3.1.6, the orifice area needs to decrease as the piston slows down. This can be accomplished with a series of holes that become covered as the piston moves over them, as shown in Figure 5.27. A tapered rod that enters a chamber in the back of the piston, as shown in Figure 5.28, grooves, as shown in Figure 5.29, or any other clever method of regulating the flow from the piston can also be used.

In addition to structural and operating considerations, the engineer needs to make sure that the cylinder can operate at the rates specified. As discussed in Section 5.4.1.5 for hydraulic motors, once the size to generate the required force has been determined, stiffness is a key characteristic driving cylinder size.

Similar to a hydraulic motor, the stiffness of a hydraulic cylinder is a function of the fluid bulk modulus, volume under compression, and the area of the piston acting on it.

On the piston side, from (5.42), that is, $\Delta V = \frac{V_o \times \Delta P}{\beta}$

$$A_p \times (\Delta x) = \frac{A_p \times x_{\text{op}} (\Delta P)}{\beta}. \tag{5.87}$$

$P = \frac{F}{A}$, thus we have

$$A_p \times (\Delta x) = \frac{A_p \times x_{\text{op}} \times \left[\Delta F_p / A_p\right]}{\beta} \tag{5.88}$$

$$\frac{A_p \times \beta}{x_{\text{op}}} = \frac{\Delta F_p}{\Delta x}. \tag{5.89}$$

As $k = \frac{\Delta F}{\Delta x}$, the piston side spring rate is

$$k_p = \frac{A_p \times \beta}{x_{\text{op}}}. \tag{5.90}$$

Likewise, on the rod side

$$k_r = \frac{A_r \times \beta}{x_{\text{or}}}. \tag{5.91}$$

By multiplying by piston or rod side area, these equations can be converted from fluid column length to volume under compression. If there is additional fluid under compression between the cylinder and control valve, it should also be added in as shown in

$$k_p = \frac{A_p^2 \times \beta}{V_{\text{op}} + V_{\text{ap}}} \tag{5.92}$$

$$k_r = \frac{A_r^2 \times \beta}{V_{\text{or}} + V_{\text{ar}}} \tag{5.93}$$

where V_{op} = piston side volume under compression inside the cylinder

V_{ap} = piston side volume under compression between the cylinder and control valve

V_{or} = rod side volume under compression inside the cylinder

V_{ap} = rod side volume under compression between the cylinder and control valve

When both the cylinder inlet and outlet are accurately controlled by a valve as shown in Figure 5.30, the stiffness of the rod and piston sides can be combined as springs in parallel.

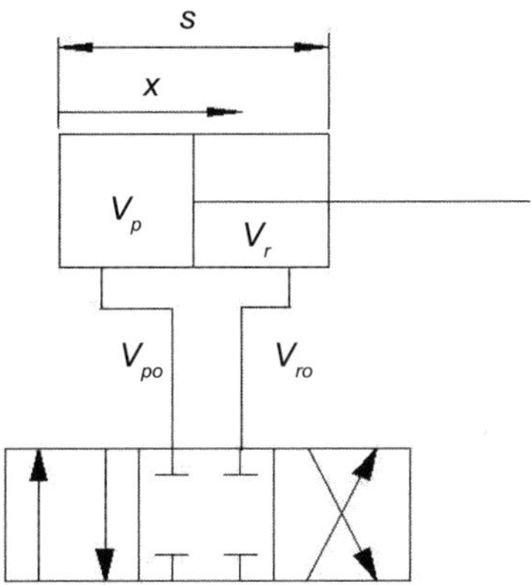

Figure 5.30 The hydraulic cylinder system.

$$k = \frac{A_p^2 \times \beta}{V_{\text{op}} + V_{\text{ap}}} + \frac{A_r^2 \times \beta}{V_{\text{or}} + V_{\text{ar}}}. \tag{5.94}$$

The natural frequency of the cylinder can be estimated by using this spring rate and the mass of the piston, rod, and reflected load, *m*, in the standard equation for natural frequency:

$$\omega_o = \sqrt{\frac{k}{m}} \tag{5.95}$$

The terminology is given as follows:

β = fluid bulk modulus

P = pressure, Pp = piston side, and Pr = rod side

V = volume

A_p = piston side area

A_r = rod side area

x = length of fluid volume under compression

x_p = piston side and x_r = rod side

V_{ap} = other volume on piston side

V_{ar} = other volume on rod side

V_{op} = initial piston side cylinder volume under compression

V_{or} = initial rod side cylinder volume under compression

m = total mass on system

k_r = rod side spring rate

k_p = piston side spring rate

- k = cylinder spring rate

$$\omega_o = \sqrt{\frac{k_p}{m} + \frac{k_r}{m}}$$

This is minimum when

$$k_p + k_r = \text{minimum}$$

or

$$\frac{A_p^2 \times \beta}{V_{\text{op}} + V_{\text{ap}}} + \frac{A_r^2 \times \beta}{V_{\text{or}} + V_{\text{ar}}} = \text{minimum}.$$

If S = cylinder stroke, $x_r = S - x_p$

$$\frac{A_p^2}{V_{\text{ap}} + A_p \times x_p} + \frac{A_r^2}{V_{\text{ar}} + A_r \times S - A_r \times x_p} = \text{minimum}$$

$$A_p^2 \times \left(V_{\text{ap}} + A_p \times x_p\right)^{-1} + A_r^2 \times \left(V_{\text{ar}} + A_r \times S - A_r \times x_p\right)^{-1} = \text{minimum}.$$

Differentiating [Swokowski 1975, 85–102]

$$\frac{d\text{minimum}}{dx} = -A_p^2 \times \left(V_{\text{ap}} + A_p \times x_p\right)^{-2} \times A_p - A_r^2 \times \left(V_{\text{ar}} + A_r \times S - A_r \times x_p\right)^{-2} \times -A_r.$$

Setting equal to 0 to find minimum x

$$0 = -A_p^3 \times \left(V_{\text{ap}} + A_p \times x_p\right)^{-2} + A_r^3 \times \left(V_{\text{ar}} + A_r \times S - A_r \times x_p\right)^{-2}$$

$$A_p^3 \times \left(V_{\text{ap}} + A_p \times x_p\right)^{-2} = A_r^3 \times \left(V_{\text{ar}} + A_r \times S - A_r \times x_p\right)^{-2}$$

$$A_p^3 \times \left(V_{\text{ar}} + A_r \times S - A_r \times x_p\right)^{2} = A_r^3 \times \left(V_{ap} + A_p \times x_p\right)^{2}.$$

Expanding, we get

$$A_p^3 \times (V_{\text{ar}}^2 + V_{\text{ar}} \times A_r \times S - V_{\text{ar}} \times A_r \times x_p + A_r \times S \times V_{\text{ar}} + A_r^2 \times S^2 - A_r^2 \times S \times x_p - A_r \times x_p \times V_{\text{ar}} - A_r^2 \times x_p \times S + A_r^2 \times x_p^2) = A_r^3 \times (V_{\text{ap}}^2 + 2 \times V_{\text{ap}} \times A_p \times x_p + A_p^2 \times x_p^2).$$

Simplifying, we get

$$\left(A_p^3 A_r^2 - A_r^3 A_p^2\right) x_p^2 - 2\left(V_{\text{ap}} A_p A_r^3 + V_{\text{ar}} A_r A_p^3 + A_r^2 A_p^3 S\right) x_p + \left(V_{\text{ar}}^2 A_p^3 + 2 V_{\text{ar}} A_r S A_p^3 + A_r{}^2 S^2 A_p^3 - A_r^3 V_{\text{ap}}^2\right) = 0.$$

Solve for x using the quadratic formula where

$$a = A_p^3 A_r^2 - A_r^3 A_p^2$$

$$b = -2\left(V_{\text{ap}} A_p A_r^3 + V_{ar} A_r A_p^3 + A_r^2 A_p^3 S\right)$$

$$c = V_{\text{ar}}^2 A_p^3 + 2 V_{\text{ar}} A_r S A_p^3 + A_r^2 S^2 A_p^3 - A_r^3 V_{\text{ap}}^2$$

and

$$x = \frac{-b \pm \sqrt{b^2 - 4ac}}{2a}.$$

As one can see from the previous sections, a high spring rate is important for high-performance actuators. For hydraulic actuators, one of the key contributors to system stiffness is the fluid bulk modulus. Typical hydraulic fluids operate with a bulk modulus of around 1378 MPa (200,000 psi) with silicone fluids generally having a lower modulus and heavy fluids such as glycerin having a higher modulus [The Lee Company 2000, M24-M28]. The engineer cannot always choose a high modulus fluid because several other factors influence hydraulic fluid selection, such as fluids in the supply chain, flammability, cost, material compatibility, and temperature range. Finally, the bulk modulus of the selected fluid is further influenced by the amount of entrained air and operating temperature. The bulk modulus drops as the amount of entrained air increases and fluid temperature rises. The effect of air in the system can be approximated by treating the air and fluid as springs in series.

From Section 5.2, we determined that the bulk modulus of air is

$$\beta = nP \tag{5.18}$$

or

$$\beta_a = \frac{c_p}{c_v} \times p \tag{5.96}$$

where β_a = bulk modulus of air

c_p = specific heat of air at constant pressure

c_v = specific hear of air at constant volume

p = air pressure

The effective bulk modulus of the fluid (β_a) is

$$\frac{1}{\beta_e} = \frac{\%\text{ Fluid by Volume}}{\beta_f} + \frac{\%\text{ Air by Volume}}{\beta_a} \tag{5.97}$$

where β_f = bulk modulus of the fluid.

From (5.97), one can see that a fluid with a bulk modulus of 1379 MPa (200,000 psi) has an effective bulk modulus of 14 MPa at sea level when there is 1% air entrained in it:

$$\frac{1}{\beta_e} = \frac{.99}{1{,}379} + \frac{.01}{\frac{2.4}{1.7} \times .1}$$

where β_e = 14 MPa (2030 psi), a 99% decrease

For a system operating pressure of 20.7 MPa (3000 psi), the effective bulk modulus is

$$\frac{1}{\beta_e} = \frac{.99}{1{,}379} + \frac{.01}{\frac{2.4}{1.7} \times 20.7}$$

where β_e = 943 MPa (137,000 psi), a 32% decrease.

At 2% entrained air, the effective bulk modulus is 713 MPa (103,000 psi) so that it is easy to see that minimizing the amount of entrained air is an important factor in maintaining a stiff system.While it is understood that the bulk modulus of a fluid decreases as its temperature increases, the amount of change is a function of fluid properties. Therefore, it is not as easy to calculate the effect of temperature changes on bulk modulus. Depending on the fluid, the bulk modulus can decrease 40% for a 50°C change. Entrained air and temperature effects on bulk modulus are cumulative, so the effective bulk modulus can change by 60% when air is present and the fluid is at high temperatures.

Table 5.1 Fluid and material properties

	Bulk Modulus		Density at 70°F		Specific Heat		Coefficient of Thermal Expansion	
Fluid	MPa	psi	kg/m³	lb/in³	kJ/kg·°C	BTU/lb$_m$°F	/°C	/°F
MIL-H-83282	2068	300,000	0.484	0.030	2.093	0.50	0.00083	0.00046
MIL-H 5606	1793	260,000	0.484	0.030	1.968	0.47	0.00083	0.00046
Water	2137	310,000	0.577	0.036	4.187	1.00	0.00038	0.00021
Silicone 100 cs	1034	150,000	0.559	0.035	1.465	0.35	0.00097	0.00054

The material property representative and actual values may be different.

Sources: The Lee Company, Technical Hydraulic Handbook, 2000, Pages M16, M20, and M25. Streeter and Wylie 1975, 711.

5.6.1.2 Pneumatic Cylinders

Pneumatic cylinders share many features and require many of the same design considerations as hydraulic cylinders. The main differences are that the fluid is less messy when it leaks, the fluid does not have good lubricating qualities, the fluid bulk modulus is much lower, and high-pressure safety concerns need to be addressed. In this section, approaches to addressing and managing the design challenges just mentioned are discussed.

Lubrication of pneumatic cylinders can be accomplished by utilizing a lubricator in the air supply, which essentially injects a lubricant (oil) into the air for lubricating air powered equipment. Drawbacks to this approach include oil consumption, oil either building up in the cylinder or being expelled as the cylinder vents to atmosphere, and oil present in the air system so that it cannot be utilized for other activities that do not require or want oil such as painting, blasting, or cleaning. These problems can be solved by utilizing seals that do not require external lubrication or incorporating a local oil lubricating system into the cylinder. Such a system, for a rod seal, can be seen in Figure 5.25. As discussed in the previous section, air has a low bulk modulus, so pneumatic cylinders have an inherent low spring rate. This low spring rate can challenge the designer of high-performance systems tying to utilize compressed air as a power source. The predominant solutions to overcoming the inherent low frequency of pneumatic systems are to increase the piston/rod side areas or add dampening to the system so that oscillations are quickly damped out. However, dampening can add challenges to achieving responsive systems. Finally, safety of high-pressure systems needs to be considered. This was discussed in Section 5.2, where a comparison between the energy stored in a hydraulic system with a high fluid bulk modulus was compared with that in a pneumatic system with a relatively low fluid bulk modulus. In the example presented there, the pneumatic system had 50 times more stored energy than the equivalent hydraulic system.

All the design challenges presented can be addressed, but the solution(s) implemented to overcome them may impact the feasibility of using pneumatics for very high performance cylinders. As a result, the engineer needs to make a comprehensive assessment of the benefits and limitations of utilizing pneumatic cylinders early in the design project's development cycle. If it is not done early in the design cycle, significant effort, time, and expense can be expended on a design approach that cannot satisfy all the design requirements.

5.6.2 Motor and Rack

One method for turning the rotary motion of a motor into linear motion is by placing a pinion on the motor shaft and letting it drive a gear rack. This design is shown in Figure 5.31.

Figure 5.31 The rack and pinion drive.

This may not seem like an elegant system, but it is just what is needed for certain applications. A relatively small motor can generate high forces, at lower velocity, if a gear reduction is placed between the motor and drive pinion. The opposite can be true if the gear reducer is replaced with a speed increaser. Rack systems can offer high stiffness but can have backlash or free play associated with them. Backlash in traditional involute profile gear form systems can be minimized by providing design provisions for adjusting the rack gear pitch line to the pinion pitch diameter. However, there will always be some backlash due to gear form errors and the desire to minimize interference (binding) between the pinion and gear rack. Particular care should be taken during design of the pinion bearing system to make sure that it has the load capacity and stiffness to control backlash, including deflection due to loads. Similar to designing or selecting a cylinder, the required motor/rack output forceand velocity can be determined as shown in Section 4.1. Once these are known, the motor and gear ratio can be calculated as discussed in Section 5.4.1.

The overall performance and efficiency of the system are a function of the components used. Efficiency and available torque can be found by multiplying the efficiency of all components in the system. For example, if the system is made up of a motor, gear reduction unit, spur gear/rack interface, and bearings, the overall system efficiency is

$$\varepsilon_{\text{Overall}} = \varepsilon_{\text{motor}} \times \varepsilon_{\text{gear reduction}} \times \varepsilon_{\text{spur gear/rack}} \times \varepsilon_{\text{bearings}}$$

where ε_x is the efficiency of component x.

Similarly, the final output force is

$$F_{\text{out}} = \frac{T_{\text{motor}} \times R_{\text{gear redution}} \times \varepsilon_{\text{Overall}}}{r}$$

where T_{motor}= motor torque, $R_{\text{gear reduction}}$ = gear reduction ratio, and r = pinion pitch radius.

Stiffness can be approximated by combining the stiffness of each component as if they are springs in series reflected to the appropriate point of reference:

$$\frac{1}{k_{\text{Overall}}} = \frac{1}{k_{\text{motor}}} + \frac{1}{k_{\text{gear reduction}}} + \frac{1}{k_{\text{spur gear/rack}}}$$

where k_x is the spring rate of component x.

In general, motor/rack systems are used for relatively short, precise, rapid motions. The motion can be multiplied by using two racks driven off the same piston in opposite directions.

5.6.3 Screws

There are several different types of screws, each offering a unique set of advantages and disadvantages. Power screws are good at turning the rotary motion of a motor into linear motion and offer a high effective gear ratio. This means the load inertia reflected back to the motor is small and a small motor can be used to move large loads. All types of screws also follow the same basic operating principles where the torque to operate then is a function the thread lead, friction between the nut, and thread and friction between screw ends and the mounting structure. Some screw types to consider include planetary roller, ball, and acme (Figure 5.32).

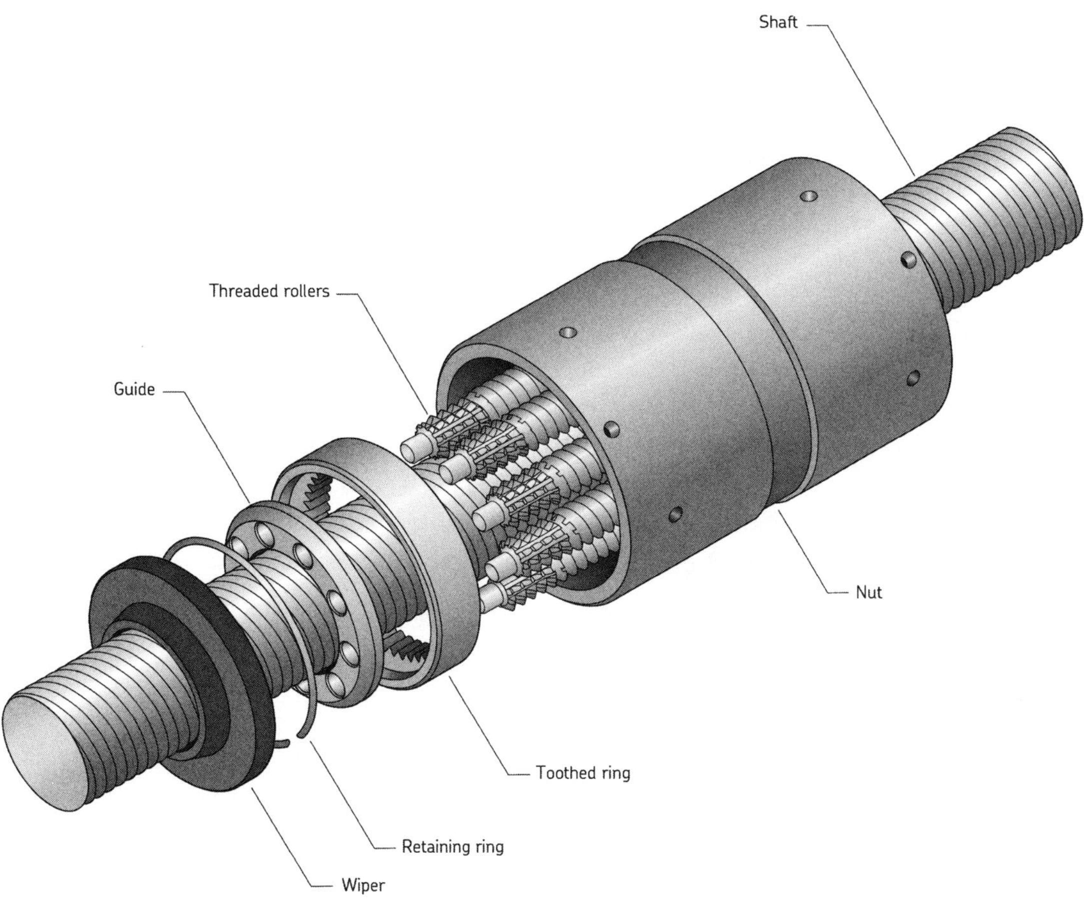

Figure 5.32 Planetary roller screw. SKF (www.skf.com). Image courtesy of SKF USA Inc.

Equations for calculating the torque to move a load can be derived from the basic geometry where

L = lead

d_t = thread pitch diameter

d_b = mean end bearing diameter

P = load

R = reaction loads at the thread (individual reaction loads are denoted by subscripts)

F = friction force (individual friction force components are denoted by subscripts)

μ_t = thread coefficient of friction

μ_b = end bearing coefficient of friction

θ = 1/2 thread form angle

α = thread lead angle = $\tan^{-1}\left[\frac{L}{\pi D}\right]$

φ = normal thread load angle ($\varphi = \tan^{-1}[\tan\theta \cos\alpha]$) (5-106)

$$T_{\mathrm{RT}} = \text{total raise torque}\left(T_{\mathrm{RT}} = P\left\{\left[\frac{\mu_t + (\cos\varphi \tan\alpha)}{\cos\varphi - \mu_t \tan\alpha}\right]\left(\frac{d_t}{2}\right) + \mu_b\left(\frac{d_b}{2}\right)\right\}\right) \tag{5.133}$$

$$T_{\mathrm{LT}} = \text{total lower torque}\left(T_{\mathrm{LT}} = P\left\{\left[\frac{\mu_t + (\cos\varphi \tan(-\alpha))}{\cos\varphi - \mu_t \tan(-\alpha)}\right]\left(\frac{d_t}{2}\right) + \mu_b\left(\frac{d_b}{2}\right)\right\}\right). \tag{5.140}$$

From a geometry perspective, the load moves a distance equal to the screw lead when the screw is turned one revolution or a distance, πD. This creates an effective ratio of $\frac{L}{\pi D}$. The friction load between the nut and screw is a function of the normal load on the threads and the coefficient of friction between the nut and threads. The thread surface is at a compound angle relative to the load. This angle is a combination of the thread lead angle and thread form angle. The thread and load transfer geometry is shown in Figure 5.33.

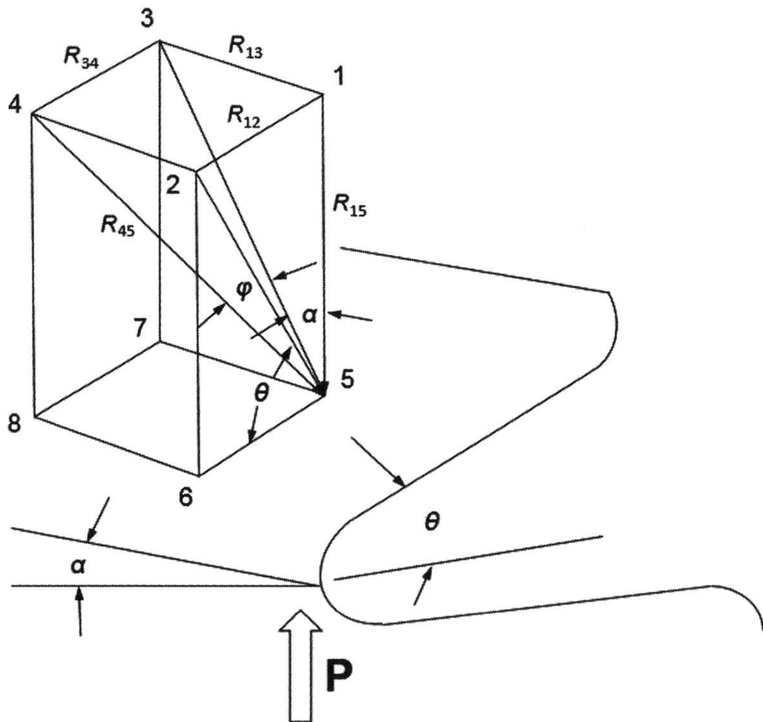

Figure 5.33 Thread geometry with load (P) and reaction forces.

From this geometry, the following relationships can be identified to develop equations relating thread friction to the torque needed to move a load P (Figures 5.34–5.36).

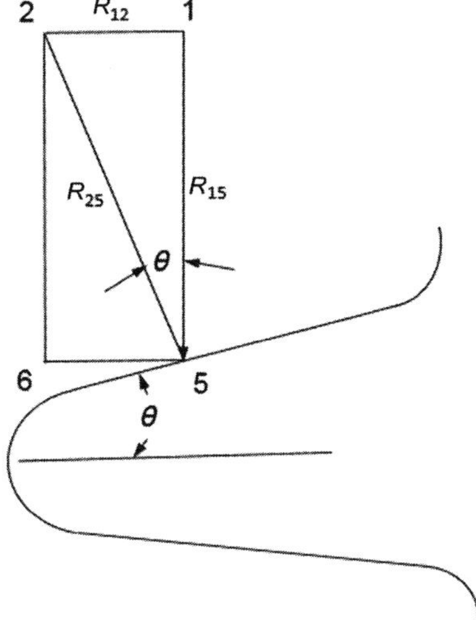

Figure 5.34 Reaction forces at thread form angle.

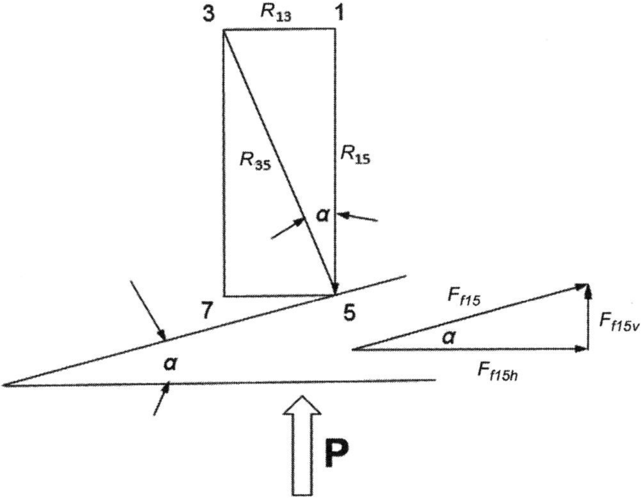

Figure 5.35 Load, thread friction, and reaction forces at thread lead angle.

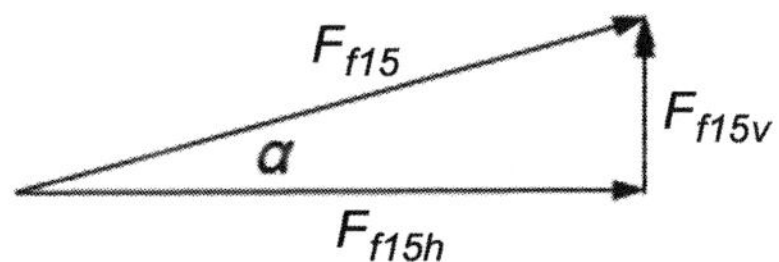

Figure 5.36 Thread friction forces.

$$R_{13} = R_{15} \tan\alpha \tag{5.98}$$

$$R_{12} = R_{15} \tan\theta \tag{5.99}$$

$$R_{35} = \frac{R_{15}}{\cos\alpha} \tag{5.100}$$

$$R_{25} = \frac{R_{15}}{\cos\theta} \tag{5.101}$$

$$R_{75} = R_{15} \tan\alpha \tag{5.102}$$

$$R_{45} = \frac{R_{35}}{\cos\varphi} = \frac{R_{15}}{\cos\alpha\cos\varphi} \tag{5.103}$$

$$\tan\varphi = \frac{R_{34}}{R_{35}} \tag{5.104}$$

$$R_{34} = R_{12}\,. \tag{5.105}$$

Combining (5.98), (5.100), (5.104), and (5.105)

$$\tan\varphi = \frac{R_{12}}{R_{35}} = \frac{R_{15}\tan\theta}{\frac{R_{15}}{\cos\alpha}} = \tan\theta\cos\alpha \tag{5.106}$$

$$F_{f15} = R_{45}\mu_t = \frac{R_{35}\mu_t}{\cos\varphi} = \frac{R_{15}\mu_t}{\cos\alpha\cos\varphi}\,. \tag{5.107}$$

The vertical component is

$$F_{f15v} = \frac{R_{15}\mu_t \sin\alpha}{\cos\alpha\cos\varphi}\,. \tag{5.108}$$

The horizontal component is

$$F_{f15h} = \frac{R_{15}\mu_t \cos\alpha}{\cos\alpha\cos\varphi}\,. \tag{5.109}$$

Summing vertical forces

$$R_{15} = P + F_{f15v}\,. \tag{5.110}$$

Combining (5.108) and (5.110)

$$R_{15} = P + \frac{R_{15}\mu_t \sin\alpha}{\cos\alpha\cos\varphi}\,. \tag{5.111}$$

Rearranging

$$P = R_{15} - \frac{R_{15}\mu_t \sin\alpha}{\cos\alpha\cos\varphi} \tag{5.112}$$

$$P = R_{15}\left(1 - \frac{\mu_t \sin\alpha}{\cos\alpha\cos\varphi}\right) \tag{5.113}$$

$$R_{15} = \frac{P}{1 - \dfrac{\mu_t \sin\alpha}{\cos\alpha\cos\varphi}}. \tag{5.114}$$

The force normal to the thread for raising a load P with a thread coefficient of friction μ_t is

$$R_{45} = \left(\frac{P}{1 - \dfrac{\mu_t \sin\alpha}{\cos\alpha\cos\varphi}}\right)\left(\frac{1}{\cos\alpha\cos\varphi}\right) \tag{5.115}$$

$$R_{45} = \frac{\text{P}}{\cos\alpha\cos\varphi - \mu_t \sin\alpha}. \tag{5.116}$$

To lower a load P, the force normal to the thread is

$$R_{45} = \frac{P}{\cos(-\alpha)\cos\varphi + \mu_t \sin(-\alpha)}. \tag{5.117}$$

For a thread pitch diameter d_t, the torque to move a load P is the sum of the horizontal forces multiplied by the thread pitch radius.

From (5.35)

$$T = \left(R_{13} + F_{\text{f15h}}\right)\frac{d_t}{2} \tag{5.118}$$

$$T = \left(R_{15}\tan\alpha + \frac{R_{15}\mu_t \cos\alpha}{\cos\alpha\cos\varphi}\right)\frac{d_t}{2} \tag{5.119}$$

$$T = \left(R_{15}\tan\alpha + \frac{R_{15}\mu_t}{\cos\varphi}\right)\frac{d_t}{2} \tag{5.120}$$

To raise a load P

$$T_R = \left(\frac{P}{1 - \dfrac{\mu_t \sin\alpha}{\cos\alpha\cos\varphi}}\right)\left(\tan\alpha + \frac{\mu_t}{\cos\varphi}\right)\left(\frac{d_t}{2}\right). \tag{5.121}$$

To lower a load P

$$T_L = \left(\frac{P}{1 - \dfrac{\mu_t \sin(-\alpha)}{\cos(-\alpha)\cos\varphi}}\right)\left(\tan(-\alpha) + \frac{\mu_t}{\cos\varphi}\right)\left(\frac{d_t}{2}\right). \tag{5.122}$$

Torque added from the end bearings is

$$T_b = (\mu_b)(P)\left(\frac{d_b}{2}\right). \tag{5.123}$$

Therefore. the total torque to raise a load P is

$$T_{\text{RT}} = T_R + T_b \tag{5.124}$$

$$T_{\text{RT}} = \left(\frac{P}{1 - \dfrac{\mu_t \sin\alpha}{\cos\alpha\cos\varphi}}\right)\left(\tan\alpha + \frac{\mu_t}{\cos\varphi}\right)\left(\frac{d_t}{2}\right) + (\mu_b)(P)\left(\frac{d_b}{2}\right). \tag{5.125}$$

The total torque to lower a load P is

$$T_{\text{LT}} = T_L + T_b$$

$$T_{\text{LT}} = \left(\frac{P}{1 - \dfrac{\mu_t \sin(-\alpha)}{\cos(-\alpha)\cos\varphi}}\right)\left(\tan(-\alpha) + \frac{\mu_t}{\cos\varphi}\right)\left(\frac{d_t}{2}\right) + (\mu_b)(p)\left(\frac{d_b}{2}\right). \tag{5.126}$$

These equations can be further simplified as follows.

For (5.125)

$$T_{\text{RT}} = \left(\frac{P}{1 - \dfrac{\mu_t \tan\alpha}{\cos\varphi}}\right)\left(\frac{\mu_t}{\cos\varphi} + \tan\alpha\right)\left(\frac{d_t}{2}\right) + \mu_b P\left(\frac{d_b}{2}\right) \tag{5.127}$$

$$T_{\text{RT}} = \left(\frac{P\mu_t}{\cos\varphi - \mu_t \tan\alpha} + \frac{P}{\dfrac{1}{\tan\alpha} - \dfrac{\mu_t}{\cos\varphi}}\right)\left(\frac{\mu_t}{\cos\varphi} + \tan\alpha\right)\left(\frac{d_t}{2}\right) + \mu_b P\left(\frac{d_b}{2}\right) \tag{5.128}$$

$$T_{\text{RT}} = P\left\{\left[\frac{\mu_t}{\cos\varphi - \mu_t \tan\alpha} + \frac{1}{\dfrac{1}{\tan\alpha} - \dfrac{\mu_t}{\cos\varphi}}\right]\left(\frac{d_t}{2}\right) + \mu_b\left(\frac{d_b}{2}\right)\right\} \tag{5.129}$$

$$T_{\text{RT}} = P\left\{\left[\frac{\dfrac{\mu_t}{\cos\varphi}}{1 - \dfrac{\mu_t \tan\alpha}{\cos\varphi}} + \frac{1}{\dfrac{1}{\tan\alpha} - \dfrac{\mu_t}{\cos\varphi}}\right]\left(\frac{d_t}{2}\right) + \mu_b\left(\frac{d_b}{2}\right)\right\} \tag{5.130}$$

$$T_{\text{RT}} = P\left\{\left[\frac{\dfrac{\mu_t}{(\cos\varphi\tan\alpha)}}{\dfrac{1}{\tan\alpha} - \dfrac{\mu_t}{\cos\varphi}} + \frac{1}{\dfrac{1}{\tan\alpha} - \dfrac{\mu_t}{\cos\varphi}}\right]\left(\frac{d_t}{2}\right) + \mu_b\left(\frac{d_b}{2}\right)\right\} \tag{5.131}$$

$$T_{RT} = P\left\{\left[\frac{\dfrac{\mu_t}{(\cos\varphi\tan\alpha)}+1}{\dfrac{1}{\tan\alpha}-\dfrac{\mu_t}{\cos\varphi}}\right]\left(\frac{d_t}{2}\right)+\mu_b\left(\frac{d_b}{2}\right)\right\} \tag{5.132}$$

$$T_{RT} = P\left\{\left[\frac{\mu_t+(\cos\varphi\tan\alpha)}{\cos\varphi-\mu_t\tan\alpha}\right]\left(\frac{d_t}{2}\right)+\mu_b\left(\frac{d_b}{2}\right)\right\} \tag{5.133}$$

Similarly, (5.126) can be simplified for lowering a load P

$$T_{LT} = \left(\frac{P}{1-\dfrac{\mu_t\tan(-\alpha)}{\cos\varphi}}\right)\left(\frac{\mu_t}{\cos\varphi}+\tan(-\alpha)\right)\left(\frac{d_t}{2}\right)+\mu_b P\left(\frac{d_b}{2}\right) \tag{5.134}$$

$$T_{LT} = \left(\frac{P\mu_t}{\cos\varphi-\mu_t\tan(-\alpha)}+\frac{P}{\dfrac{1}{\tan(-\alpha)}+\dfrac{\mu_t}{\cos\varphi}}\right)\left(\frac{d_t}{2}\right)+\mu_b P\left(\frac{d_b}{2}\right) \tag{5.135}$$

$$T_{LT} = P\left\{\left[\frac{\mu_t}{\cos\varphi-\mu_t\tan(-\alpha)}+\frac{1}{\dfrac{1}{\tan(-\alpha)}+\dfrac{\mu_t}{\cos\varphi}}\right]\left(\frac{d_t}{2}\right)+\mu_b\left(\frac{d_b}{2}\right)\right\} \tag{5.136}$$

$$T_{LT} = P\left\{\left[\frac{\dfrac{\mu_t}{\cos\varphi}}{1-\dfrac{\mu_t\tan(-\alpha)}{\cos\varphi}}+\frac{1}{\dfrac{1}{\tan(-\alpha)}+\dfrac{\mu_t}{\cos\varphi}}\right]\left(\frac{d_t}{2}\right)+\mu_b\left(\frac{d_b}{2}\right)\right\} \tag{5.137}$$

$$T_{LT} = P\left\{\left[\frac{\dfrac{\mu_t}{(\cos\varphi\tan(-\alpha))}}{\dfrac{1}{\tan(-\alpha)}-\dfrac{\mu_t}{\cos\varphi}}+\frac{1}{\dfrac{1}{\tan(-\alpha)}+\dfrac{\mu_t}{\cos\varphi}}\right]\left(\frac{d_t}{2}\right)+\mu_b\left(\frac{d_b}{2}\right)\right\} \tag{5.138}$$

$$T_{LT} = P\left\{\left[\frac{\dfrac{\mu_t}{(\cos\varphi\tan(-\alpha))}+1}{\dfrac{1}{\tan(-\alpha)}-\dfrac{\mu_t}{\cos\varphi}}\right]\left(\frac{d_t}{2}\right)+\mu_b\left(\frac{d_b}{2}\right)\right\} \tag{5.139}$$

$$T_{LT} = P\left\{\left[\frac{\mu_t+(\cos\varphi\tan(-\alpha))}{\cos\varphi-\mu_t\tan(-\alpha)}\right]\left(\frac{d_t}{2}\right)+\mu_b\left(\frac{d_b}{2}\right)\right\}. \tag{5.140}$$

When T_{LT} is negative, the load is overdriving the screw and an opposing torque, represented by the negative value, must be applied to hold the load. The transition to back-driving the screw is at $T_{LT} = 0$. From (5.140), we see that

$$\mu_b\left(\frac{d_b}{d_t}\right) = \left[\frac{\mu_t+(\cos\varphi\tan(-\alpha))}{\mu_t\tan(-\alpha)-\cos\varphi}\right]. \tag{5.141}$$

If bearing friction is ignored, (5.141) becomes

$$0=\left[\frac{\mu_t+(\cos\varphi\tan(-\alpha))}{\mu_t\tan(-\alpha)-\cos\varphi}\right] \tag{5.142}$$

$$-\mu_t=(\cos\varphi\tan(-\alpha)) \tag{5.143}$$

$$\mu_t=\cos\varphi\tan\alpha. \tag{5.144}$$

[Spotts 1978, 249–254]

For actuators utilizing screws, the torque to raise (T_{RT}) or lower (T_{LT}) a load can be calculated based on the equations derived above. Determining appropriate values for μ_t and μ_b can be challenging, but screw and bearing manufacturers often provide values for their products. As discussed previously, stiffness is also an important factor for high-performance actuators.

As part of power screw design and selection, the design engineer needs to consider screw buckling for the loads over the range of operation. Screw loads may vary with position due to geometry, unbalance, and payload acceleration/deceleration. This analysis along with other stiffness and strength requirements helps define screw size and bearings. The bucking capacity of a screw can be estimated by using the screw root diameter in the buckling calculations shown in Section 5.6.

5.6.4 Spring Rate for Power Screw Actuators

The overall spring rate for a screw drive is made up of all the components in the system. Typical components include the mounting structure, screw end bearing, screw, nut, and load structure. The system spring rate is the result of the individual spring rates combined as a set of springs in series

$$\frac{1}{k_{\mathrm{ST}}}=\frac{1}{k_{\mathrm{MS}}}+\frac{1}{k_B}+\frac{1}{k_S}+\frac{1}{k_N}+\frac{1}{k_{\mathrm{LS}}} \tag{5.145}$$

where k_{ST} = total screw drive system spring rate (force/change in length)

k_{MS} = mounting structure spring rate (force/change in length)

k_B = screw end bearing spring rate (force/change in length)

k_S = screw spring rate (force/change in length)

k_N = screw nut spring rate (force/change in length)

k_{LS} = load structure spring rate (force/change in length)

5.6.4.1 Mounting Structure and Load Structure Stiffness

The spring rate for the mounting (k_{MS}) and load structures (k_{LS}) can be calculated from the spring rate of the components or extracted from finite element analysis of the structure. The mounting structure is defined as the structure that directly mounts the screw and nut hardware. The spring rate for the mounting structure (k_{MS}) can be calculated from a deflection analysis of the mounting components. Initial calculations may want to simplify them and beams or members in tension or compression so that simple engineering deflection calculations can be performed. As the design is refined, the spring rate can be calculated from a more sophisticated "hand calculation" or finite element analysis of the components. The load structure (k_{LS}) is defined as the larger structure that the mounting structure is connected to and forms the load path between the screw and/or nut and "ground." Determining the spring rate of the load structure is utilized the same initial and refinement approaches as the mounting structure.

5.6.4.2 Screw Stiffness

Stiffness for the screw (k_S) can be obtained from manufacturer's data or calculated as if the screw is treated as a column under load. The spring rate for a component under tension or compression can be determined as follows:

E = elastic modulus of the material (force/area)

σ = stress (force/area)

ε = strain (change in length/length)

k = spring rate (force/change in length)

$$E = \frac{\sigma}{\epsilon} \tag{5.146}$$

$$\sigma = \frac{F}{A} \tag{5.147}$$

$$\varepsilon = \frac{\Delta l}{l} \tag{5.148}$$

$$E = \frac{F \times l}{A \times \Delta l} \tag{5.149}$$

$$k = \frac{F}{\Delta l} \tag{5.150}$$

$$E = k \times \frac{l}{A} \tag{5.151}$$

$$k = \frac{\text{EA}}{l} \text{ or for the screw, } k_S = \frac{EA}{l}. \tag{5.152}$$

5.6.4.3 Nut Stiffness

The best place to get stiffness of the nut (k_N) is from the screw and nut manufacturer's data. If it is not available from a supplier, it can be calculated by treating the contact between the screw thread and nut thread as two cantilevered beams. This requires the thread form to be known. For an acme thread, the general form is shown in Figure 5.37.

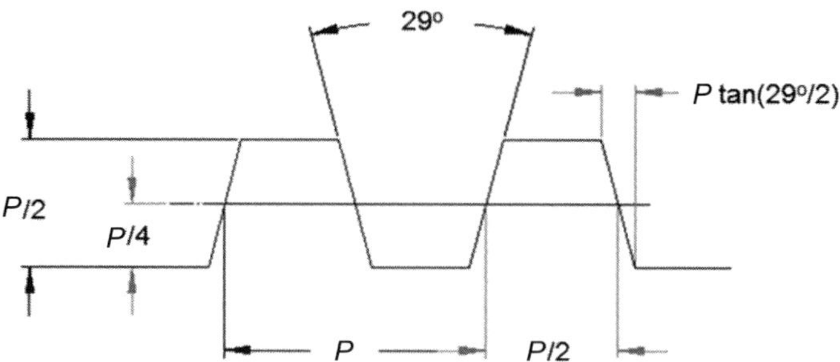

Figure 5.37 Acme Thread Form, where P = thread pitch, L_b = size of the thread base [$P/2$ + ($P/2$) tan(29°)], and L_t = size of the thread tip [$P/2$ – ($P/2$) tan(29°)].

Ideal interaction between the screw and nut takes place along the pitch line, which is $P/4$ from the base of the thread for both the screw and nut. Knowing that the deflection of a cantilever beam is $\delta = \frac{Fl^3}{3EI}$, deflection of the screw thread and nut thread can be calculated:

where F = load

L = length of the beam or in this case $P/4$

E = elastic modulus of the material

I = area moment of inertia of the thread or $I = \frac{bh^3}{12}$

where b = length of the thread contact between the screw and nut and h is conservatively approximated by $h = L_t + (2/3)(L_b - L_t)$, knowing that contact between the nut and screw is not perfect.

Total deflection at the nut comes from defection of the screw thread and nut thread. The deflection of the nut thread is

$$\delta_n = \frac{Fl^3}{3E_n I}. \tag{5.153}$$

The deflection of the screw thread is

$$\delta_s = \frac{Fl^3}{3E_s I} \tag{5.154}$$

where E_n is the elastic modulus for the nut material and E_s is the elastic modulus for the screw material. The total deflection is

$$\delta_T = \frac{Fl^3}{3E_n I} + \frac{Fl^3}{3E_s I}. \tag{5.155}$$

5.6.4.4 Total Screw/Nut Stiffness

The spring rate for the screw–nut combination is $k_N = \frac{F}{\delta_T}$:

$$k_N = \left[\frac{3I}{l^3}\right]\left[\frac{1}{\left(\frac{1}{E_n}+\frac{1}{E_s}\right)}\right]. \tag{5.156}$$

5.6.4.5 Bearings

As seen in (5.145), screw end bearing stiffness (k_B) also plays an important role in the overall stiffness of a screw drive system. Proper bearing selection depends on the structure, screw interface, and stiffness requirement. Popular bearing types include angular contact ball bearings, spherical roller bearings, tapered roller bearings, and cross roller bearings. When determining the appropriate type and size bearing to select, the dynamic load, static load, speed profile, and stiffness should be considered. As with other bearing applications, the load capacity of the design can be improved by using multiple bearings, but the appropriate scaling factors need to be used. It is best to get these factors from the bearing manufacturer because it is dependent on bearing type, materials, and manufacturing tolerances. For example, the dynamic load capacity for two bearings may be 1.5 times that of a single bearing and 2.2 times single bearing capacity for three bearings. However, for the same bearing, the static load may be linear with the number of bearings, that is, two bearings have twice the static capacity as a single bearing and three bearings have three times the static capacity as a single bearing. Selecting the bearing type depends on cost, space available, and structural stiffness. Cross roller bearings may be compact but need a stiff supporting structure. Tapered roller bearings offer good radial load capacity, but their axial load capacity is lower. Tapered roller bearings offer the ability to be preloaded to minimize or eliminate backlash at the bearing and improve stiffness characteristics. Spherical roller bearings offer good radial and axial load capacities with the added advantage of being less sensitive to misalignment. They can be preloaded to minimize or eliminate backlash and improve stiffness but, for similar size bearings, do not have stiffness levels comparable to tapered roller bearings. Angular contact ball bearings are offered in specific configurations for screw end bearings. They offer good radial and axial load capacities, the ability to be preloaded for low or no backlash, and high stiffness. The bearing sets can be arranged for equal load and stiffness capacities in both extend and retract directions or tuned to have higher ratings in one direction. Depending on the specific design requirement, one of these bearings, other rolling element radial/thrust bearing combinations, or one of the many journal bearing types may be appropriate for the application. The engineer needs to assess load capacity (radial and axial), life, environment, operating characteristics (friction, backlash, and stiffness), and cost against the requirements as part of the final bearing selection process.

5.6.5 Screw Critical Speed

Another item to be understood is the critical speed of the screw. This is the speed at which the screw may start experience vibration due to the speed being near one of the screw modes. This can be estimated by calculating the natural frequency of the screw. Considering the screw to be a cylinder with a diameter equal to the pitch diameter and knowing that the greatest deflection occurs at the center of the screw span, the natural frequency is estimated based on a concentrated load at mid span deflection of the screw at mid span. As with any beam, screw deflection depends on screw end conditions and length:

$$k_s = \frac{P}{\delta} \tag{5.157}$$

where k_s = screw spring rate, P = load, and δ = deflection at load P

$$f_n = \sqrt{\frac{k_s}{m}} \tag{5.158}$$

where fn = screw natural frequency (radians/second), k_s = screw spring rate, and m = applied mass of the screw and nut [Beer and Johnston 1976, 747]

or

$$f_{nrpm} = f_n \times \frac{\text{r}}{2\pi \text{ rad}} \times \frac{60 \text{ s}}{\text{min}} \tag{5.159}$$

where f_{nrpm} is the natural frequency of the screw in rpm.

For example, for a simply supported 10-mm-diameter screw, 2.5 m long, and a combined effective screw/nut mass of 1.5 kg:

$$\delta = \frac{Pl^3}{48EI}$$

$$I = \frac{\pi d^4}{64}$$

$$I = \frac{\pi (10)^4}{64}$$

$$I = 491$$

$$\delta = \frac{P \times (2500)^3}{48 \times 206843 \times 491}$$

$$\delta = 3.2 \times P$$

$$k_s = \frac{P}{\delta}$$

$$k_s = \frac{P}{3.2 \times P}$$

$$k_s = .31$$

$$f_n = \sqrt{\frac{.31}{(1.5/1000)}}$$

$$f_n = 14.4$$

$$f_{nrpm} = 14.4 \times \frac{1}{2\pi} \times \frac{60}{1}$$

$$f_{nrpm} = 138$$

If one wants to operate at 75% of the natural frequency, then the system should be designed to $f_{nrpm} = .75 \times 138$ and $f_{nrpm} = 100$ rpm.

5.6.6 Electric Solenoids

Another way to accomplish linear motion, particularly small motions, is with electric solenoids. Electric solenoids do not have high-performance drive characteristics because simple solenoids have two positions, extended or retracted. When energized, they move from one position to the other, depending on whether the solenoid is extended or retracted in the unenergized position. However, solenoids can be useful in high-performance drive systems as support components to drive simple devices such as latches. Therefore, they are covered briefly in this section.

Electric solenoids are relatively simple devices composed of an electric coil with a plunger. When the coil is energized, the magnetic field created causes the plunger to move (Figures 5.38 and 5.39). Using electric solenoids can be a simple, inexpensive method to move short distances quickly. However, they do not offer the ability to follow a specific acceleration, velocity, or deceleration profile. Acceleration is controlled by the force generated by the solenoid at the beginning of the motion and the mass of the load $\left(a = \frac{F}{m}\right)$. Magnetic flux density for a cylindrical solenoid is defined by Ampere's law, as shown in

$$Bl = \mu NI \ . \tag{5.160}$$

where B = magnetic flux density (T)

l = solenoid length (m)

μ = magnetic constant of $4\pi \times 10^{-7}$ (T·m/A)

N = number of wire turns in the coil

I = current (A)

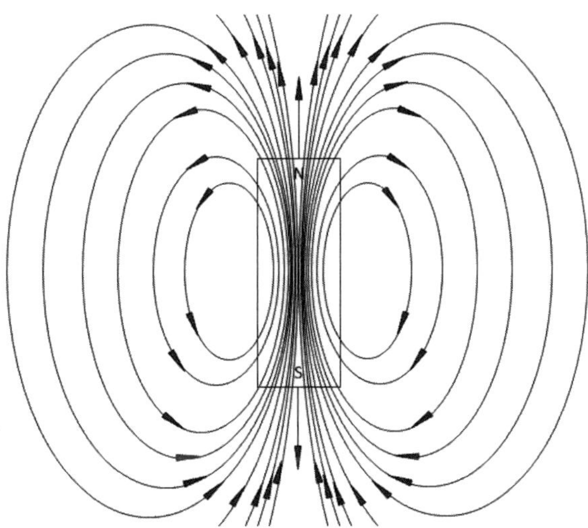

Figure 5.38 General solenoid magnetic field.

Figure 5.39 The electric solenoid—the TLX Technologies long stroke latching solenoid (www.tlxtech.com). Image courtesy of TLX Technologies.

If coil end effects are neglected, the total magnetic flux through a coil is obtained by multiplying the flux density by the cross-sectional area (*A*)

$$B_T = \left[\frac{\mu NI}{l}\right] \times A\,. \tag{5.161}$$

Knowing that the inductance is

$$L = \frac{NB_T}{I} \tag{5.162}$$

the inductance for a solenoid can be determined to be

$$L = \frac{\mu N^2 A}{l}\,. \tag{5.163}$$

The energy stored by an inductor is

$$E_P = \frac{I^2}{2} L\,. \tag{5.164}$$

[Shortley and Williams 1971, 606–608, 659–663]

Similarly, one can differentiate mechanical potential energy with respect to the change in length to get force

$$E_P = \frac{1}{2}kx^2 \quad \frac{dE_P}{dx} = F = kx \tag{5.165}$$

If the solenoid is approximated as an electromagnet with a uniform magnetic field, one can differentiate the energy stored by an inductor, in this case a solenoid, with respect to the change in length to get the force as it moves. First, (5.163) and (5.164) are combined

$$E_P = \frac{I^2}{2}\left[\frac{\mu N^2 A}{l}\right] \tag{5.166}$$

$$\frac{dE_P}{dl} = F = -\frac{I^2}{2}\left[\frac{\mu N^2 A}{l^2}\right] \tag{5.167}$$

or

$$F = -\frac{\mu I^2 N^2 A}{2l^2} \tag{5.168}$$

where l = solenoid movement, gap, or stroke.

From this equation, one can see that the force decrease rapidly as solenoid stroke increases. The acceleration portion of an actuator cycle generally occurs when the gap is the greatest, that is, the plunger is extended from the coil and when the solenoid can generate its lowest force. This can result in a solenoid that is oversized for the remaining portion of the move (maintaining constant velocity and deceleration). The only resistance to motion offered by an electric solenoid comes from friction. Therefore, for the deceleration portion of a move cycle, solenoids do not offer significant resistance to motion, so another device, such as a shock absorber, is required for a controlled stop. Adding components can increase complexity and any passive devices do not offer the ability to utilize information from feedback devices to improve performance.

Electric solenoids are best suited for moving small loads short distances. In these applications, they can offer the ability to rapidly move the load by simply turning current ON and OFF. The force/energy/power available from electric solenoids can be amplified by using them to control the flow of another media such as air or hydraulic fluid. Their simplicity, low cost, and reliability make them a good choice for remotely controlling large systems.

Proportional solenoids are electric solenoids where armature force is determined by an electric control signal (Figure 5.40). As they are solenoids, their behavior is governed by Ampere's law, but the design of the solenoid components, including the pole(s), is refined so that they can move to intermediate points between the end positions. The addition of a control system, drive electronics, and refined internal components makes the overall system more complex and expensive. However, incorporating proportional solenoids into a design may improve capability and performance at minimal cost. It all depends on overall system requirements, manufacturing and supply chain approach, and engineering expertise available.

Figure 5.40 The proportional electric solenoid—the TLX Technologies proportional solenoid (www.tlxtech.com). Image courtesy of TLX Technologies.

As mentioned earlier, some benefits of solenoids are their simplicity and therefore cost, reliability, and availability. They are good for moving relatively small loads at short distances rapidly. However, in their simplest form, they do not have the ability to follow a controlled motion profile and are subject to getting hot if energized for long periods of time. They generally need to be energized to hold a load in position, but the current creating the magnetic field in the solenoid is also creating heat when no work is being done. This temperature rise can affect solenoid operation because an increase in temperature increases resistance, which has the undesired effect of reducing the solenoid force available.

5.6.7 Electric Linear Motors

Electric linear motors are gaining popularity as actuators for linear motion (Figure 5.41). The same motion control fundamentals such as stiffness, output force, energy efficiency, and integration into the overall system space claim are important. The high-performance actuator engineer needs to evaluate the linear motors available that can best meet all design requirements and create a system from them. Early in the design process, the different types of linear motors need to be assessed against the requirements along with their advantages and disadvantages.

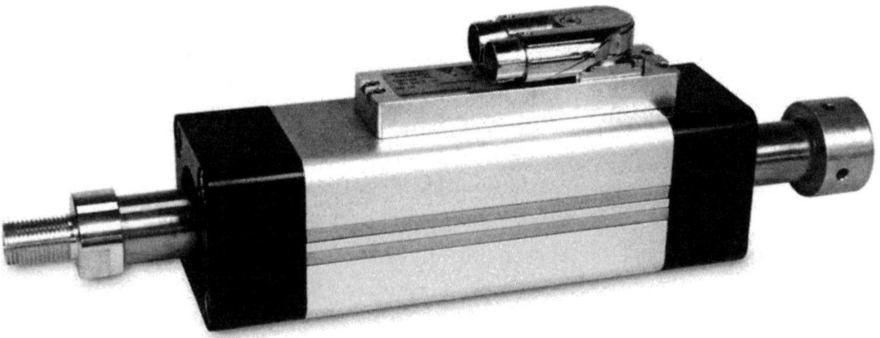

Figure 5.41 The linear electric motor—the ETT Series Parker electric tubular motor (www.parker.com/eme/ett). Image courtesy of Parker Hannifin Electromechanical & Drives Division.

Overall, advantages of linear motors include low mass for the dynamic portion of the system, high positioning accuracy, and low noise. They can easily change direction and can have a long service life due to a few components and low friction. On the other hand, linear motors may not be able to generate the force required for large applications or to achieve very high acceleration rates. The load is directly coupled to the drive so that the load inertia reflected back to the motor is not reduced, the heat generated due to motor inefficiency is directly at the actuator and therefore load, and the new technologies introduce unknown risks into a design. Constant improvements in linear motor technology continue to increase the number of applications for linear motors.

Looking at how the linear motor advantages affect the approach to the actuator design, we see that the low dynamic mass of the motor can improve system response through improved acceleration and deceleration rates. For weight and energy conscious designs, reduced dynamic mass can have a large system impact due to reduced structural requirements, smaller drive components, including power management and control systems. With high-positioning, accuracy at the load, which is closely coupled to the motor, has the benefit of achieving the required load positioning accuracy with a fewer components, added system components, or complexity needed to compensate for load positioning error. For systems with low-noise requirements, a drive with inherent low-noise generation reduces system, size, weight, complexity, and cost by eliminating the need for acoustic management features. Such features can include sound insulation, structural tuning for system natural frequencies, and active noise cancelling systems. The ability of linear motors to change direction of travel for the load rapidly within the motor with minimal components between the motor and load allows for a robust design and low wear. Again direct coupling between the motor and load means that there is no backlash in intermediate drive components that can wear and/or impact. These wear and impact points may mean increased system maintenance to both prolong component life and replace components as they wear out. Wear and/or impact can also contribute the overall system noise just discussed. The ability to have a low number of support features in a system designed around linear motors along with the type of components used may allow the dynamic friction points to be lower than for other types of designs. Again this can have multiple benefits including lower overall energy requirements, minimizing losses that go to heat, reducing maintenance to compensate for friction induced wear, and reducing system complexity needed to accommodate maintenance and/or component replacement.

Actuator design can also be affected by the need to compensate for the weaknesses of linear motors. If a single linear motor cannot generate the amount of force needed to accelerate, move, hold, or adequately control the load, it is not a candidate for the design without modification. The engineer has several choices, but from a linear motor perspective, there are only three that are significant. Performance requirements can be changed to match the force that the linear motor is capable of producing a different motor can be selected or the design can be altered to accommodate multiple linear motors. For the case where load inertia is too large compared with motor inertia for satisfactory control, there are three major choices: increase motor inertia or decrease load inertia until the desired load to motor inertia ratio is achieved, decrease performance requirements until the ratio is satisfactory, or change the ratio between the motor and load to achieve the desired load inertia seen by the motor. If reflected load inertia is a significant factor in the design, using a smaller motor with high speed (and travel) can be the most efficient approach because force and speed change linearly with the motor/load ratio, while the inertia changes with the square of the motor/load ratio as shown in (3.114) and (3.115). Heat generated due to motor inefficiency can contribute to several design challenges. These include selecting materials for the high-temperature areas that can maintain the required mechanical properties, making sure that the load in not exposed to temperatures that are above its limit for safety, durability, or function, and the design is tolerant of dimensional changes caused by thermal expansion or contraction. Examples of design accommodations that may need to be made for temperature include utilizing high-performance "plastics," using metals or ceramics, and high-temperature antifriction coatings to reduce wear, improve coefficient of friction characteristics, and manage galling. If the load is sensitive to high temperatures, either passive or active cooling may need to be incorporated into the design. The complexity of a cooling system may mitigate system cost, reliability, and/or performance benefits gained by using a linear motor. Making a high-performance actuator tolerant to thermal expansion/contraction can be relatively easy if the design is straight forward and similar materials are used throughout the design so that the thermal expansion values change only by component size and temperature differential. However, if the design is complex, utilizing several materials with widely varying coefficients of thermal expansion in addition to the component size and temperature differential effects, the task of accommodating heat from a linear motor can become more challenging. Such designs need to be evaluated carefully with proper attention given to the time it takes for development and testing to minimize risks associated with the increased complexity. The same can be said for all new technologies or new application(s) of existing technologies.

Utilizing linear motors in a high-performance drive may be the path to a new performance level, but as with all design decisions, the pros and cons of the choice need to be carefully assessed and documented. Completing this step early in the design process allows the justification for choosing a linear motor to be easily re-evaluated as complications are encountered as the design matures. This is a natural part of the design process and should not be viewed as unique to linear motors. [Aerotech Inc. 2010, 5-27]

5.6.8 References

Beer, F. P., and Johnston Jr., E. R. 1976. Mechanics for Engineers, Dynamics Third Edition. McGraw-Hill Book Company: New York.

Byars, E. F., and Snyder, R. D. 1975. Engineering Mechanics of Deformable Bodies, Third Edition. Intext Educational Publishers: New York.

John Deere Service Publications. 1979. Hydraulics, Fundamentals of Service. Deere & Company: Moline, IL.

Shortley, G., and Williams, D. 1971. Elements of Physics, Fifth Edition. Prentice-Hall, Inc: Englewood Cliffs, NJ.

Sperry Vickers. 1970. Industrial Hydraulics Manual. Sperry Rand Corporation: Troy, MI.

Spotts, M. F. 1978. Design of Machine Elements, Fifth Edition. Prentice-Hall, Inc: Englewood Cliffs, NJ.

Swokowski, E. W. 1975. Calculus With Analytic Geometry. Prindle, Weber & Schmidt, Inc: Boston, MA.

The Lee Company Technical Center. 2000. Technical Hydraulic Handbook. The Lee Company: Westbrook, CT.

Aerotech Inc. 2010. Linear Motors Application Guide. Aerotech, Inc: Pittsburgh, PA.

Streeter, V. L., and Wylie, E. B. 1975. Fluid Mechanics. McGraw-Hill Book Company: New York, NY.

5.7 Rotary Actuator Design

When high-performance rotary motion is required, there are three primary types of actuators that can be used: direct drive motors, motors with gear boxes, motor and/or linear drive with a mechanical mechanism, and rack driving a gear. As with linear actuators, the engineer can select from several basic designs within these types based on power source, performance requirements, design life, and cost.

5.7.1 Direct Drive Motors

Using a motor to directly drive a load can provide the simplest design solution to moving the load from one position to another. The engineer needs to be sure that the items just covered for linear motors such as adequate torque, positioning accuracy, ability to change direction, service life, and ability to control with the reflected load inertia are considered during motor selection. Design considerations for motor selection are covered in Section 5.4.1.

5.7.2 Gear Boxes

Gear boxes are a key component in rotary actuator design. They are used to match the torque and speed requirements of load with the motor to create the smallest, lightest, and most efficient package possible. This section is not about how to design a specific gear box, where there are many other good resources for that, but rather how to determine the characteristics needed for an application so that the gear box can be designed or selected. The key characteristics of a gear boxe include ratio, backlash, stiffness, static and dynamic load capacity, duty cycle capability, shaft load, and overall package configuration/size.

Determining the ratio for a gear box includes understanding the output torque and speed requirements for the load and the input torque and speed capability of the drive motor. [Spotts 1978, 468–469] Determining the output torque and load requirements was covered in Section 3.2 and understanding the drive motor capability was covered in Section 5.4.1. Other factors that influence the ratio include the inertia of the motor and the load inertia reflected back to the motor. Ultimately, this ratio is determined by the control system, but ratios greater than 10:1 require more complexity in the control algorithm with ratios in the range of 5:1–10:1 being even more desirable. Reflected inertia is changed by the square of the gear ratio, so the desired gear ratio for known load and motor inertias can be calculated as follows:

$$N_I = \frac{I_R}{I_m} \tag{5.169}$$

$$I_R = \frac{I_L}{R^2} \tag{5.170}$$

$$N_I = \frac{I_L}{R^2 \times I_m} \tag{5.171}$$

where I_L = load inertia

I_m = motor inertia

I_R = reflected load inertia

N_I = gear ratio between the load and motor for inertia

The gear ratio needed to match motor torque with the torque required for load acceleration, friction, and unbalance is simply the ratio shown in

$$N_T = \frac{T_m}{T_a + T_f + T_{\text{ub}}} \quad (5.172)$$

where T_m = motor torque

T_a = torque required for acceleration at the load

T_f = torque required to overcome friction at the load

T_{ub}= torque required to overcome unbalance at the load

Likewise, the gear ratio needed to match motor speed with the maximum velocity required by the load is shown in

$$N_S = \frac{V_L}{S_m} \quad (5.173)$$

where S_m = maximum motor speed and V_L = maximum load velocity.

Final gear box ratio selection needs to accommodate the requirements of N_I, N_T, and N_S. [Hulst and Kamstra 2012, 10]

Backlash in a gear box affects the ability of the control system to follow the desired motion profile and achieve the positioning accuracy desired with stable control. Physically, backlash is the gap between gear teeth and it shows up as lost motion between the motor and load. When the motor is driving, backlash shows up as motor motion without any movement of the load and when the load is back driving the motor, as during deceleration, it shows up as load motion without any corresponding motor motion or load-induced torque. For high-performance actuators, high backlash values can be problematic as the controller tries to make the load adhere to the motion profile with little position, velocity, or acceleration error. It can become even more pronounced if the load reverses during the motion cycle. Obviously, low backlash is good from a control viewpoint, but producing a gear box with low backlash can be expensive and difficult as tolerances for each gear must be held tight, shafts and shaft hole position and size must be controlled, and bearings must be selected to achieve the total tolerance required. Depending on the approach to reduce backlash, very low backlash gear boxes may also not be as efficient as others with more backlash if there are more friction losses between components because there is little clearance or even slight interference between them—at least during certain portions of the operating range. Some gear box types lend themselves to achieve low backlash better than others. Bevel gear types can adjust the position of the pinion or mating ring gear to minimize backlash, and spur gear boxes need to have eccentrics designed into the shaft locations to allow gear centers to be adjusted to minimize backlash. However, in these gear boxes, backlash can only be reduced to the point of zero clearance between the tightest teeth of mating gears. In other gear box types such as planetary gearboxes that have multiple gears in contact at the same time, backlash can be managed through the statistical clearance stackup between all the mating gears. In all cases, gear quality affects backlash with higher quality gears providing the opportunity to minimize backlash in the system. Gear quality is generally defined by specifying gear quality in terms of an American Gear Manufacturers Association quality number. The total composite tolerance associated with a given gear quality number is influenced by gear diametral pitch and pitch diameter with large gears have larger tolerances than smaller gears. Higher quality numbers represent higher quality gears. By evaluating and controlling all the gear interface dimensions listed previously, one can determine and control the amount of backlash in a gear system. [Spotts 1978, 461–462]

The effect of stiffness on drive performance is similar to backlash except that it is always present, not only when the gap between gears is present. The effect of stiffness on drive performance was discussed in Sections 1.8, 3.1.6, and 3.3; it basically affects the overall natural frequency of the drive through the following relationship:

$$f_n = \sqrt{\frac{k}{I}} \tag{5.174}$$

where k = the spring rate (system stiffness) and I = mass moment of inertia.

Stiffness is also an indicator of how much energy may be stored in the gear box. This stored energy needs to be managed by the control system as the gear box consumes energy during acceleration/deceleration phases of the motion profile and gives it up during the constant velocity phase(s) of the profile. This follows the general potential energy equation:

$$E_p = \frac{1}{2}k\theta^2 \tag{5.175}$$

where k = the spring rate (system stiffness) and θ = gear box angular deflection. [Shortley and Williams 1971, 153]

The static and dynamic capability of a gear box defines its ability to handle the loads imposed on it. The engineer need to fully understand the static and dynamic loads that the actuator experiences during operation and rest to adequately specify or design a gear box. Depending on the application, static loads may be greater than the dynamic loads. In many cases, they are. Even though the gear box may not be moving, it may experience that many static load cycles over its life if peak static loads are generated during transportation or other external factors. These static loads along with their frequency need to be calculated and compared with gear box capability. The dynamic loads come from the actuator cycle and are dependent on the motor's peak and rated torque values along with the commanded torque values needed to achieve the specified motion profile. Gear box components need to be strong enough to handle these loads for the specified life of the actuator.

The duty cycle along with corresponding torque values need to be known, so a gear box with adequate life can be selected. For static loads, component strength is important and the load frequency just needs to be known so that a gear box with adequate life for the application can be specified. For dynamic loads, strength needs to be considered, so adequate life is achieved, but the ability of the gear box to handle thermal loads also needs to be considered. Thermal loads are a function of gear box efficiency and the frequency of the load. For example, a 90% efficient gear box that has an average input torque of 10 N·m of torque during a move needs to handle 1 N·m of energy per cycle. If there is only 1 cycle/min, this may not be significant, but if there is a cycle every 0.1 s, it may be. The ability to handle these thermal loads may impact the overall gear box package size, configuration, and integration into the entire actuator.

How actuator loads are imposed on the gear box input and output shafts are also an important consideration. Actuator loads on gear box shafts may include radial, moment, and axial loads. The engineer needs to be clear on the loads and types of loads to make sure that the actuator is capable of handling them. When a gear box has a bearing that only supports radial loads, any moment or thrust loads transferred to the shaft, load the gear box in ways that is was not designed to handle. This can cause poor performance, shortened service life, or in extreme cases immediate gear box failure. The same situation can apply at the input shaft and motor. If the shaft is not designed to accommodate the loads transferred to it, the following options are available:

1. Select a different gear box that can handle the loads.
2. Redesign the gear box to handle the loads.
3. Redesign the mating shaft end connection so the loads transferred to the gear box are within the capability of the gear box.

One must also be aware of any misalignment between the motor and gear box input shaft, and the gear box output shaft and load. If these are rigid connections that can impose loads on either component, adding a coupling to limit transferred radial and/or axial loads should be considered.

The final area to be considered for actuator gear boxes is their packaging. It needs to align with actuator geometry and fit in the overall actuator space claim while satisfying structural and thermal requirements. As discussed previously, the gear box needs to accommodate the structural requirements to manage backlash, stiffness, and transmitted loads. To satisfy these requirements, the housing/overall package needs to be designed for the gear type selected, which in turn may be influenced by the desired input/output shaft orientation; that is, in line, parallel, or right angle. Housing strength and stiffness need to meet bearing and assembly needs to also satisfy backlash and drive stiffness needs. Thermal loads are a function of overall gear box efficiency and duty cycle. If efficiency is high and duty cycle low, the gear box may not need to dissipate significant thermal energy. However, if efficiency is low and/or duty cycle high, thermal loads may be significant and the housing design would need to accommodate dissipation of the thermal energy. This may be done through increased lubricant capacity, fins to promote heat transfer, or in extreme cases active cooling of the gear box through direct lubricant cooling or circulating an external coolant through the gear box.

Avoiding gear boxes in high-performance actuator design can be desirable to reduce complexity and cost. However, as discussed in this section, incorporating gear boxes in a design may offer the benefit of matching drive motor capability with load requirements while also enhancing controllability. If this is done effectively, the final actuator design using a gear box can meet all actuator requirements.

5.7.3 Mechanical Mechanisms

The function of any actuator is to move a load between positions. Sometimes this task is best performed with a mechanism that moves the load through the desired path and can be controlled to achieve the move with an acceptable motion profile. The advantage with mechanisms is that they can be optimized for a given application. However, it can also be a disadvantage when actuator production volumes are low and the engineering and manufacturing tooling costs cannot be amortized over a large number of parts. As with all actuator design activities, mechanism design needs to include understanding the basic requirements, determining the appropriate drive to load ratio, backlash, stiffness, static and dynamic load capacity, duty cycle capability, and overall package configuration/size.

In mechanism design, engineers can utilize their knowledge of basic math and trigonometry to write equations of motion for the mechanism and then use calculus to derive the equations for velocity, acceleration, and jerk. These can be used to determine the torque and speed requirements for designing the mechanism. The force/torque profile needs to be fully understood to determine the prime mover power requirements. Total power requirements need to consider efficiency of the mechanism along with any other supporting components such as gear box, linear drive, and/or motor. One of the advantages of mechanical mechanisms is that they can be tuned to offer a variable drive ratio. Ideally, this has the highest motor to load ratio at the beginning and end of the load cycle so that the greatest force/torque is available for acceleration/deceleration at the beginning of the motion profile and the highest speed is available during the constant velocity portion of the profile.

For control purposes, the overall backlash for the mechanism needs to be calculated or at the very least estimated. Overall backlash comes from clearance at pivot points and clearance in sliding slots. The design of pivot points and sliding slots depends on the basic mechanism concept, design life in number of cycles, radial loads, thrust loads, reversing loads, speed/velocity, and ambient environment. Location of pivot and/or slot locations is defined as part of developing the mechanism concept and writing the equations of motion. Other interface, clearance requirements, drive connections, and load connections define the rest of the mechanism. Within these constraints, space needs to be allocated for the pivot pins or shafts and any associated bearings. The pins or shafts need to be designed to manage applied loads and satisfy stiffness requirements. Considered loads should include pin shear and bending loads along with direct or contact loads. These loads need to be evaluated for both pin/shaft and bearing size. Space for bearings may be a factor in bearing selection with rolling element bearings generally taking up more space than plain bearings. The load and speed rating of the bearing need to match those of the pivot. In general, the pressure times velocity (*PV*) rating of the bearing should be verified against the maximum applied values at each connection. When determining overall backlash, the clearance at each pivot needs to be reflected back to the location of interest (input or output) by multiplying it by the ratio between the pivot and location of interest.

Mechanism stiffness is also of interest as it determines the structural natural frequency of the mechanism and influences controllability. In a mechanism designed for a specific application, stiffness can also be optimized within the limits of the design space. As in all structural components, the stiffness of a mechanism's components is a function of the geometry and material:

$$k = f\left(I_A, E, x\right)$$

where I_A = area moment of inertia of the component, E = elastic modulus for the component, and x = load position relative to component supports.

The mass of a mechanism is important because it is related to the frequency through $f = \sqrt{k/m}$ and contributes to the force or torque required to achieve the desired performance through $F = ma$ or $T = I\alpha$.

The loads and their behavior need to be fully understood though the entire motion profile to determine the size of mechanism components for both strength and stiffness.

Mechanism design can be rewarding, but the engineers new to the field should make sure that they fully understand the requirements, including the operating environment for the mechanism. Thermal growth and contraction need to be considered if the mechanism has to operate in large temperature extremes. Contamination and how to manage it need to be considered for most mechanisms because a few operate in a truly clean environment. Corrosion, both from the external environment and galvanic between mechanism components, needs to be considered for the same reason along with wear. Finally, galling needs to be considered and steps taken to avoid galling at dynamic interfaces of the mechanism.

5.7.4 Racks

Racks were discussed in Section 5.6.2 for linear actuators. They can also be used to turn linear motion into rotary motion. Using racks to generate rotary motion can be advantageous when the application needs to rotate a fix amount in one direction and then return to its original position. As with other drive types, the effective ratio, backlash, stiffness, and integration into the overall design are the important design parameters.

The ratio for this type of drive is

$$N = \frac{\text{Rack Displacement}}{\text{Load Displacement}}$$

As linear drives can be capable of generating a large force and it may be desirable to have the load move more that the rack, it is possible for N to be less than 1. This may be beneficial from a load/force viewpoint, but the engineer needs to make sure it is also acceptable from a reflected inertial viewpoint. As

$$I_R = \frac{I_L}{R^2} \tag{3.115}$$

the reflected inertia can be greater than the load inertia, making the load harder to control, especially in high-performance applications. For backlash, low ratios can also be problematic because any backlash near the drive connection is not reduced at the load. On the positive side, this type of drive usually has a few sources of backlash—just at the connection between the linear actuator and rack and between the rack and pinion. For these systems, drive stiffness tends to be high because rack is stiff, mainly encountering tensile and compressive loads where stiffness is a function of the material elastic modulus, loaded cross sectional area, and loaded length:

$$k_r = \frac{E \times A}{l}$$

where k_r = rack spring rate

E = material elastic modulus

A = loaded area

l = loaded length

The remaining items contributing to drive stiffness are rack and pinion teeth and output shaft stiffness. Rack and pinion stiffness is dependent on the gear form and size selected for the application. It may be approximated in a manner similar to that for thread stiffness shown in Section 5.6.3

$$\varphi = \frac{L \times T}{J \times G}$$

where φ = shaft defection

T = applied torque

L = shaft length

J = shaft polar moment of inertia (for a solid round shaft, $J = \dfrac{\pi \times d^4}{32}$

G = material shear modulus

Integration of a linearly driven rack to create rotary motion may be easy or challenging, depending on the rest of the design. In fact, it may become one of the significant factors in the decision to use a rack or some other drive for a rotary application.

5.7.5 References

Spotts, M. F. 1978. Design of Machine Elements, Fifth Edition. Prentice-Hall, Inc: Englewood Cliffs, NJ.

Shortley, G., and Williams, D. 1971. Elements of Physics, Fifth Edition. Prentice-Hall, Inc: Englewood Cliffs, NJ.

Hulst, S., and Kamstra, L. 2012. Gearmotors: Achieving the Perfect Motor & Gearbox Match. Groschopp, Inc: Sioux City, IA.

5.8 Feedback Systems

There are two types of control systems, open loop and closed loop. An actuator with open loop control is told to go somewhere and then expected to go there without correction. With closed-loop control, the actuator is told to go somewhere and feedback from a sensor provides information on how well the actuator is following orders. If the error between where it appears to be going and where it has been told to go is too great, then corrections are made to reduce the error. Actuators with open-loop control systems are less complex and easier to keep stable than ones using closed-loop control. However, addition of sensors and increased control algorithm complexity in closed-loop systems can compensate for a less robust mechanical design. It is generally good practice to design an actuator with robust mechanical control characteristics to allow control flexibility during final algorithm tuning and testing. Mechanical design of feedback systems for an actuator includes sensor type and sensor location.

5.8.1 Sensor Type

There are several types of sensors that may be useful to help control an actuator. Characteristics to consider when selecting a feedback device include interface to the control system, environment (shock, vibration, temperature range, etc.), accuracy, need to maintain absolute position information, mounting requirements, and cost. [Hulst and Kamstra 2012, 15] Each of these items and the combination of them determine the best feedback device for a given application. The more common types of sensors include position, velocity, acceleration, pressure, and temperature. Within each of these groups, there are subgroups from which the engineer can select sensors that meet the mechanical and control requirements.

For mechanical actuators, position sensors are utilized frequently. Depending on the accuracy required and control complexity employed, information from this sensor may be differentiated to get velocity and acceleration. Similar information can also be obtained from velocity and acceleration sensors, except that their information may be integrated if position and/or velocity information is needed. There are two classes of position sensors, rotary and linear. Velocity and acceleration sensors can be rotary, linear, or inertial.

5.8.1.1 Rotary Sensors

Rotary sensors measure displacement in angular coordinates. If a rotary sensor is mechanically connected to a linear actuator, it can provide linear position, velocity and/or acceleration information. When extremely high precision is needed or absolute position information needs to be retained, a multispeed position sensor system can be employed. In a multispeed system, multiple sensors operating at different speeds or ratios to the load are utilized. For absolute position information, one sensor operates such that one revolution of the sensor is equal to or less than full travel of the load. Other sensors may operate at a higher speed to provide additional position, velocity, or acceleration resolution. As sensor inertia is low, geared, and multispeed sensors generally utilize antibacklash gears to eliminate backlash from the feedback system. Determining how many sensors are required and at what speed they should operate needs to be done in collaboration with control and electronics engineers to make sure that the correct information is available for control and the sensor integrates properly with the system power and electronic design approach.

Determining which type of rotary position feedback device to use depends on the application. Some popular rotary feedback devices include optical encoders, resolvers, and tachometers. These can all be easily tailored to use in specific applications.

Optical encoders generally provide a digital signal that is defined by a beam shining through a clear disc with radial lines that break the beam to provide information at the specified interval (Figure 5.42). This type of device primarily provides velocity information that must be integrated to get position or differentiated to get acceleration. As the exact position of the disc is not defined, the position of the actuator needs to be calculated and stored in the control system memory. However, as the information comes in digital format, encoders integrate easily into modern digital control systems.

Figure 5.42 Rotary Optical Encoder. Turck RS-24/RS-31 Optical Encoder (www.turck.com). Image courtesy of Turck.

Resolvers are a type of rotary transformer that provides an analog signal specific to position of the rotor to the stator. For two-pole resolvers, resolver output is equal to the mechanical position of the rotor relative to the stator, that is, the electric angle is equal to the mechanical angle. Multipole resolvers can be used in application where greater accuracy is needed. In multipole resolvers, the electrical angle is equal to the mechanical angle times the number of pole pairs. Information from resolvers is processed through resolver to digital (R–D) converters to transform the sine and cosine resolver information into binary information to use in digital control systems.

Tachometers provide angular speed information when an independent source of speed information is desired. As sources of position information become more robust and the ability to rapidly differentiate this with velocity and acceleration information, the need for velocity specific feedback decreases. Today most tachometer information is available in digital format for easy integration with digital control systems. [Ohshima et al. 1988, 59–176, 225–243]

5.8.1.2 Linear Sensors

Linear sensors measure displacement in linear coordinates. If a linear sensor is mechanically connected to a rotary actuator, it can provide angular position, velocity and/or acceleration information. Determining linear sensor type and integration needs to be done in collaboration with control and electronics engineers to make sure that the correct information is available for control and the sensor integrates properly with the system power and electronic design approach.

Determining which type of linear position feedback device to use depends on the application. Some popular linear feedback devices include linear variable displacement transformers (LVDTs), magnetoresistive sensors, magnetostrictive sensors, and potentiometers (Figure 5.44). Integration of these sensors into the actuator package may be straight forward for short-to-medium stroke application but may be harder for actuators with long strokes in harsh environments.

Figure 5.43 LVDT sensor—the Trans-Tek Series 500 LVDT (transtekinc.com). Image courtesy of Trans-Tek, Inc.

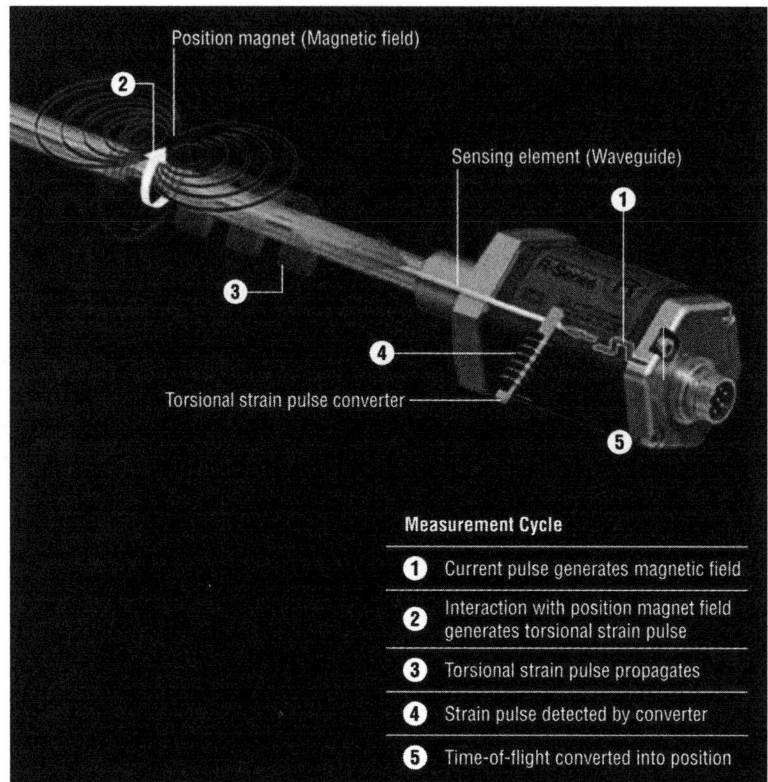

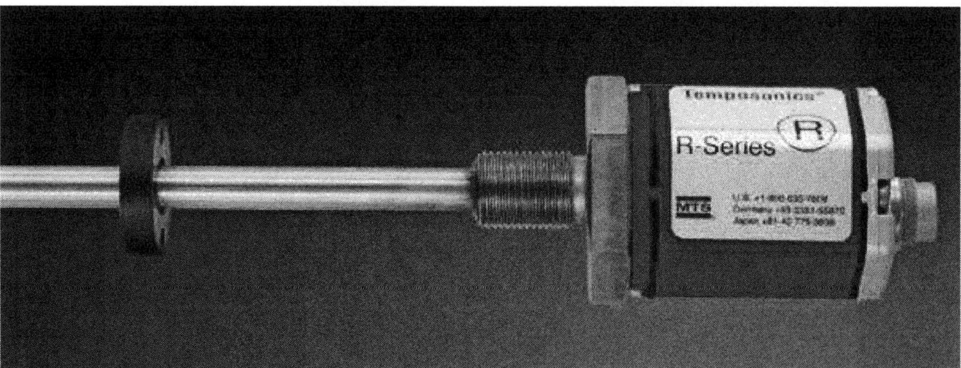

Figure 5.44 Magnetostrictive linear position sensor—the MTS R-Series sensor (www.mtssensors.com). Image courtesy of MTS Systems—Sensor Division.

LVDTs rely on electromagnetic coupling to supply position information. Position comes from moving a core along the axis of a tube that has three coils assembled end to end around it. As the core moves, the resulting voltage phase and amplitude indicate position and direction. LVDTs are used in many applications due to their robust nature, tolerance to harsh environments including temperature extremes, and ability to retain position information when power is removed. [Shortley and Williams 1971, 646–649]

Magnetoresistive sensors utilize the magnetoresistive effect to provide position information. The magnetoresistive effect is the resistance of a ferromagnetic material carrying a current due to the presence of a magnetic field. The location of an external magnetic field (magnet) relative to the ferromagnetic material carrying the current creates unique resistance value that can be translated into position information. Magnetoresistive sensors are robust and retain absolute position information. However, they can be sensitive to temperature changes and other external magnetic fields, which may also change the resistance of the ferromagnetic material.

Magnetostrictive sensors have a ferromagnetic measuring element and a position magnet that moves relative to this element and generates direct axis magnetic field in the element. As a current pulse passes through the element, a radial magnetic field is created around the element. The interaction between direct axis magnetic field and the radial magnetic field generates a strain pulse that travels from the intersection of these two fields to the end of the element where it is detected. The position of the radial magnetic field is determined by measuring the time that elapses between when the current pulse is released and when the strain pulse is detected at the end of the element. Magnetostrictive sensors are robust and retain absolute position information, even after power has been lost. As there should not be contact between the magnet and ferromagnetic measuring element sensor wear is not a factor. When implementing these sensors, the engineer needs to make sure that temperature effects and external magnetic fields cannot influence the measurement to create unacceptable error. [Ohshima et al. 1988, 181–221, 225–243], [MTS Sensors 2014, vi]

5.8.1.3 Inertial Sensors

Inertial sensors essentially utilize a mass on a spring to indicate acceleration through measuring spring displacement (Figure 5.45). The sensors can be very small, utilizing piezoelectric, piezo resistance, or capacitance changes to measure the displacement and provide the output signal. These devices can be internally configured to measure acceleration in one, two, or all the three axes of motion. Again, these sensors are robust but do not have the ability to retain absolute position information.

Figure 5.45 The inertial sensor—the Honeywell MAQ13 accelerometer (www.honeywell.com/sensing). Image courtesy of Honeywell Sensing and Control.

Other feedback devices that may be desired to improve controllability or position accuracy include pressure and temperature sensors. Pressure sensors may be used more frequently in hydraulic and pneumatic applications. One can choose from two categories of pressure sensor: absolute and differential (Figures 5.46 and 5.47). Absolute pressure sensors provide the absolute pressure value for the monitored location, while a differential sensor provides the pressure difference between monitored locations. [Haslam, Summers and Williams 1981, 177–178] Differential pressure can be calculated from two absolute sensors but may not be as accurate because the inaccuracy of each sensor is added to get the differential reading, while a differential sensor utilizes one sensor to compare the measured pressure values with each other. Temperature sensors may also be utilized to compensate for mechanical system or other sensor information changes due an actuator operating in and environment with a large temperature range (Figure 5.48).

Figure 5.46 The pressure sensor—the Turck PT1 Pressure Sensor (www.turck.com). Image courtesy of Turck.

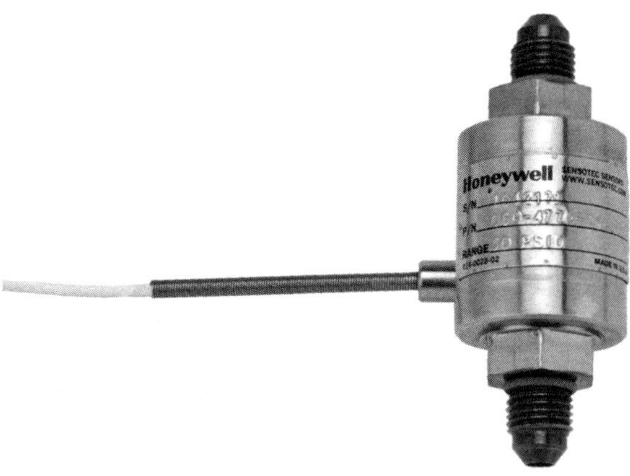

Figure 5.47 The differential pressure sensor—the Honeywell AS25D Pressure Sensor (www.honeywell.com)/sensing. Image courtesy of Honeywell Sensing and Control.

Figure 5.48 The temperature sensor—the Turck TTM temperature sensor (www.turck.com). Image courtesy of Turck.

5.8.2 Sensor Location

Location of the feedback sensors is significant to performance of the actuator. As stated earlier, open-loop systems do not provide adjustments to actuator position or error based on a feedback sensor, while closed-loop systems use load position to reduce the error. If a feedback sensor is located on the load, it can provide load information within sensor accuracy. As the sensor mounting is moved further from the load and closer to the prime mover, knowledge of the load is reduced due to inherent differences in the mechanical system. However, position errors from the mechanical components can also cause instability in the actuator as the control algorithm tries to follow the motion profile and locate the stopping point. Some of the characteristics associated with actuator motion profile and behavior are discussed in Section 3.1.6. Stiffness of the actuator also influences the best location for position feedback devices. If the sensor is being excited by its location, the resulting command to the prime mover may be influenced in a way that amplifies the excitation and causes erratic actuator performance.

Sensors should also be located to avoid unnecessary environmental effects. Some other factors that could have adverse effects on sensor performance include temperature, vibration, moisture, dirt, and physical loads. Sensors should be placed so they operate within their linear temperature range, so ambient temperature, external heat loads, and heat generated from the actuator (inefficiencies) need to be considered. Most sensors also have a specified frequency range within which they are certified to operate. During actuator design, all sensor locations should be evaluated to make sure that they are not excited to a level beyond the sensors capability. Some sensors are hermitically sealed so that moisture and dirt do not affect their performance, life, or reliability, but others do not. Any sensors that are not capable of surviving the contamination that they are exposed to need additional shielding designed into their actuator interface. If additional shielding is added, the engineer needs to verify that the shield does not adversely affect sensor performance, or temperature or vibration loads. Finally, any external physical loads that may be placed on sensors need to be considered and the effects mitigated. Many sensors are placed in locations that provide convenient hand holds or steps. Most sensors and associated cables cannot survive the pulling, pushing, or other point loads that may be associated with equipment use, inspection, or service. If a sensor is located where it may be used as a step or hand hold, additional protection should be provided. This protection should also be evaluated to make sure that it does not affect sensor performance or create detrimental temperature or vibration loads.

5.8.3 Actuator Components to Stop/Hold a Load

A key characteristic of high-performance actuators it that they can rapidly go to a position with little error and hold that position if necessary. Motion profiles that can meet the performance requirements for a given application are covered in Section 3.1.6. As discussed in Section 3.1.6, move times can be shortened significantly if velocity does not go to zero at the end of the move but stops from a predetermined velocity. It may also be desirable to securely hold the load in the designated position. Two approaches to accomplish these functions are to use a brake or latch.

Brakes can be used to stop and hold a load as commanded. There are two basic types of brakes: friction and tooth. Friction brakes utilize long life friction surfaces to stop and/or hold a load. They can be configured to engage when force is applied or removed. Most applications desire to have the brake engage when power is removed for safety reasons. For this configuration, springs cause the brake to engage bringing the load to a stop if commanded or power fails. The load is then securely held until power is again applied to release the brake. Friction brakes can be an easy solution except that they can be large for a given application and may not easily integrate into the space available for the actuator (Figure 5.49).

Figure 5.49 Friction brake—the Carlyle Johnson MAXITORQ FEA fail-safe brake (cjmco.com). Image courtesy of Carlyle Johnson.

A solution to a smaller brake can be to use a multidisc brake or utilize a tooth brake. Tooth brakes can be an efficient means to hold a stopped load because in addition to friction, they also use a mechanical means to hold the load (Figure 5.50). However, because they utilize mechanical interference to help hold the load, they are not good for stopping a moving load.

Carlyle Johnson MAXITORQ JEB Jaw Brake (cjmco.com)

Figure 5.50 Tooth brake—the Carlyle Johnson MAXITORQ JEB jaw brake (cjmco.com). Image courtesy of Carlyle Johnson.

Latches can be similar in principle to a tooth brake, but rather than multiple teeth engaging the load, a single latch engages a slot or hole to hold a load in position (Figure 5.51). They can also be used to stop a load from an acceptable velocity at the end of a move cycle. This approach supports a smooth deceleration and stopping profile, as defined in (3.56), $\frac{a}{v} \leq C\omega_n$ or $\frac{dv}{dx} \leq C\omega_n$, where one can see that reaching zero velocity can take an infinite amount of time.

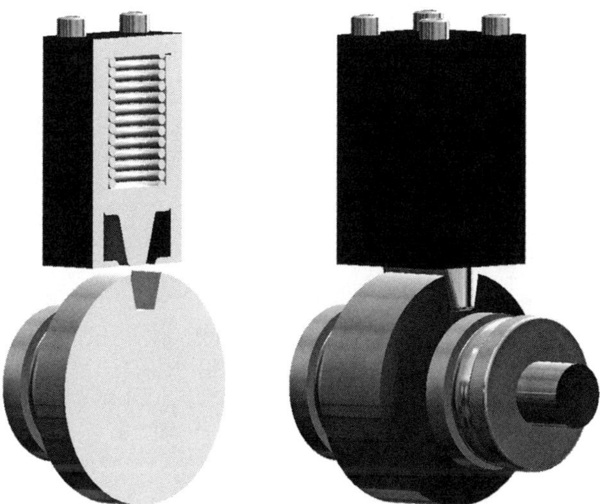

Figure 5.51 Latches.

Both brakes and latches can be used for linear and rotary actuators. The load they must hold is a function of their location in the actuator assembly. If they are mounted on or at the prime mover, they only have to manage the load as seen there, which may be reduced by the effective speed reduction. Placing a latch or brake at the prime mover location has the advantage of reducing the load and therefore reducing its size at the price of the brake or latch load being at a higher speed. Also, as the brake or latch is placed further from the load, the risk of losing control of the load due to component failures between the latch/brake location and the load increases. Therefore, selecting, designing, and locating brakes/latches in an actuator become a series of compromises among cycle time, brake/latch size, and risk to loss of load control. [Spotts 1978, 279–296]

5.8.4 References

Spotts, M. F. 1978. Design of Machine Elements, Fifth Edition. Prentice-Hall, Inc: Englewood Cliffs, NJ.

Shortley, G., and Williams, D. 1971. Elements of Physics, Fifth Edition. Prentice-Hall, Inc: Englewood Cliffs, NJ.

Hulst, S., and Kamstra, L. 2012. Gearmotors: Achieving the Perfect Motor & Gearbox Match. Groschopp, Inc: Sioux City, IA.

Ohshima, Y., Akiyama, Y., Hata, K., Himemo, H., Masaki, K., Mori, H. et al. 1988. Servo Sensors, Elements and Applications. Intertec Communications Inc: Ventura, CA.

Haslam, J. A., Summers, G. R., and Williams, D. 1981. Engineering Instrumentation and Control. Edward Arnold Ltd: London.

MTS Sensors. 2014. Temposonics(R) Magnetostrictive, Linear-Position Sensors. MTS Sensors: Cary, NC.

5.9 Verification and Validation

In the design to requirements phase of a project, verification and validation revolve around the specific analysis techniques and tools used for this specific project that reduce project risk, building confidence in the actuator's ability to satisfy customer requirements. Validation of the analytical models and tools used to predict the mechanical behavior of the actuator system should occur prior to or early in this phase so that confidence in the result they predict is understood and steps can be made to mitigate risks for areas where confidence is lower. Tool and model validation should be based on an evolution of the activities performed during the requirements analysis phase. During this phase, there should be more time and commitment to allow test results to be obtained, as needed, to improve confidence in the verification techniques identified during and after requirements analysis. Analysis techniques with proven correlation to test results provide confidence that the actuator can satisfy the defined goals during the verification and validation phase. At this point, any areas that show risk in meeting the goals should have risk mitigation plans in place.

Chapter 6
Prototyping

Building and testing prototypes is a large part of any development project, especially one where new technology is utilized. A project can employ several stages of prototype build and test depending on the maturity of the design and technology incorporated into the design. Early in the development stage, component level or subassembly level prototypes may be built to verify the basic design approach and verify that the technology is ready to be released in a product for general use. After the initial design is complete, prototypes should be built to verify the manufacturing strategy, provide platforms for verification testing, and gather data from user testing. Knowledge learned from these activities can then be incorporated into a final design that can be built with production tooling, go for additional user testing, and then be released for production and commercial sale. While this appears to be a simple linear process, in reality, it may take several designs, analysis, and build and test iterations before the design is ready for commercial sale. Prototype build–test iterations can be expensive to every effort should be made to ensure requirements are accurately and completely defined early in the development process and development follows good engineering process steps, including analysis and cross-functional team input. Making and testing prototypes should be both a rewarding and learning experience for all involved in the development process.

6.1 Fabrication

Prototype fabrication can take place in three stages: technology development, design verification and validation, and production readiness. The purpose of each phase is different, but all should have the goal of learning about the design and leading to production for commercial sate. After all, commercial sale of the design pays for the development and provides funding for future improvements.

In the technology development stage, prototypes to verify that the technology is ready for use in the actuator design are made. For high-performance actuators, this stage may have various activities including procurement of materials to support component manufacture, component manufacture to the specifications needed for the application, and procurement of the necessary components from suppliers. At this stage, the development team needs to make sure that the materials specified by the design are available in both the composition and configuration desired. If not, the design needs to be altered to use materials that are available. Sometimes, material availability is not really known until it is requested. Once raw material is available, the next step is to actually make the required components to the shape, size, and accuracy required for the application. At this time, the engineer also becomes aware of any design requirements that are not adequately specified or are over specified in the documentation (drawings). Inadequate definition of design requirements can lead to parts that meet documentation (drawings) requirements but do not meet the intent of the design and cannot perform or fit as required. Overspecification of design requirements, such unnecessarily tight tolerances, can make a part impossible or costly to make. High-performance actuators may have components that need to be procured from a supplier that specializes in them. Such components may include motors, sensor, gear reduction units, linear and rotary bearings, and fasteners. While these may be procured from a supplier, they also may be customized for the application. At the technology development stage, identification of potential suppliers should to be made and components procured from them to make sure that they can be obtained to the specifications needed. Parts manufactured at this stage are used to verify that the technology is ready to be included in a production intent design.

The preliminary design verification stage requires complete prototypes to be developed so that they can be tested against the design requirements. The same categories of material procurement, component manufacture, and procurement of supplier components need to be completed. However, at this stage of the development and fabrication cycle, the details of manufacturing each component need to be considered. Items to be considered at this time include cost, lead time, supplier relationship, tooling, and ability to get parts as specified. By the time a design has reached this level of development, cost targets should be established for the product and these decomposed to cost targets for each component. This information can be used along with estimates for actual part procurement to develop an understanding of how well the design is going to meet the cost targets without additional cost reduction efforts. The lead time or time to get a part from the time it is ordered becomes definitively known when orders are placed and parts actually received. A design may include some long lead time items and these should be identified as early as possible. This allows documentation for these items to be completed and orders placed early in the fabrication/procurement cycle so that all components are available to support assembly and test functions. During this stage, prototype suppliers for raw materials, custom parts, modified parts, and catalog items should be finalized. Part specifications should be reviewed during the design process, but additional reviews with suppliers should take place just before the actual orders are placed to make sure that the suppliers can deliver the parts as specified and any tooling required to make the parts is identified. Parts manufactured at this stage are used to verify that the design can be made and meets the design requirements.

At the production readiness stage, the ability of the actuator to meet the design requirements has been verified. The production plan to manufacture the design at a competitive cost and in quantities needed to satisfy customer demand must be created and implemented. Based on previous manufacturing, assembly, and test information, changes may be incorporated that enhance manufacturability. These changes may take the form of changing manufacturing techniques, such as moving from a welded or machined component to a cast component. Such changes may require minor changes to the design or design documentation. Knowledge about component tolerance requirements may also be gained during assembly and testing of earlier prototypes that allow changes to be made to enhance manufacturability. As the manufacturing plan matures, production tooling can be designed, procured, and incorporated into the manufacturing process. Production methods with this tooling may also require minor adjustments to the design or documentation to facilitate its use. As more prototype parts are manufactured, the ability to apply statistical process controls and understand the manufacturing capability may also influence the final design and documentation.

As fabrication of parts at each stage is completed, the parts become available for assembly and test. Assembly and testing of these parts influences the maturation process of fabrication, and likewise, part fabrication influences the maturity of the assembly and test stages. Together they work to mature the final actuator design.

6.2 Assembly

As with design and fabrication, prototype assembly can happen in three stages: technology development, design, and production readiness. The level of assembly difficulty should decrease as the design matures at each stage. The level of participation by the design organization should decrease and involvement of the manufacturing organization's assembly team should increase as product development progresses through the stages.

Assembly of the first prototype generally requires participation from the engineer(s) and designer(s). Involvement of the design organization accomplishes two objectives: any assembly questions or concerns can be immediately answered and the design staff sees all assembly issues first hand. The objective of a technology development prototype is to verify that the technology utilized in the design is ready for customer use from both a design and manufacturing perspective. If it is lacking in either area, additional development is needed. This may affect how the development program is classified. A technology development prototype that displays design stability issues relative to the fundamental design requirements and/or manufacturing issues may need to be classified as research and development (R&D) rather than product design. Clearly understanding the state of the technology that needs to be incorporated into the design for it to meet the design requirements is the goal of the technology demonstrator prototype.

When the technology is deemed to be at a satisfactory maturity level, the design and part fabrication moves to the preliminary design assembly phase. At this point, the organization should be comfortable that the technology is stable enough to support the product's basic design requirements. Knowing this, the fabricated parts should be managed jointly by the engineering and manufacturing organizations. Engineering needs to supply design expertise and guidance because all design, fabrication, assembly, and test documentations are not complete at this stage. Completing an understanding of documentation needs is part of this effort. Manufacturing needs to supply expertise on assembly techniques and fixtures. Ultimately, the manufacturing organization is going to be responsible for producing the product for sale so that assembly of the preliminary design prototypes is their first opportunity to work with the entire design and provide feedback that should be incorporated into the production readiness prototype. This is also manufacturing's opportunity to identify additional assembly fixtures, training, documentation, and subsystem quality assurance activities needed. Utilizing information from this stage, the final design can be completed and parts for the production readiness prototype fabricated.

Assembly of the production readiness prototype should be done with little assistance from the design organization. All assembly fixtures and documentation should be in place. Personnel performing the assembly tasks should have received training identified during the preliminary design prototype phase. Quality assurance testing and personnel should be involved to ensure that the product meets the specifications and requirements defined in manufacturing/assembly documentation. There should be several production readiness prototypes built for long-term testing and initial testing with selected customers. From a design and manufacturing perspective, this should be the final step with a few changes made before production starts.

6.3 Prototype Verification and Validation

During the Prototype phase, verification and validation of the manufacturing process starts. It is here that the engineer gets the first glimpse of how well the actuator, as designed, can be manufactured. It includes procurement of raw materials, components from suppliers, in-house manufacturing capability, and process steps such as heat treatment, plating, and other steps that require unique equipment or capabilities that may be provided by outside suppliers. Verification determines that the design can be produced to the specifications provided in the engineering documentation such as drawings, interface documents, and written specifications. Validation determines if all the actuator can be assembled and performs as expected when everything produced complies with the engineering documentation.

Chapter 7
Verification and Validation

Prototype verification and validation follows the same staging with similar objectives as fabrication and assembly. Product verification and validation testing should be done by the product verification/validation and quality organizations to ensure an independent and nonbiased assessment of the design. At this point, we should also be clear on the difference between verification and validation. A product is verified when it is tested against the defined requirements and is deemed to meet them. A product is validated when it successfully operates as intended through the range of conditions as expected by the user/customer. The goal of requirements analysis and decomposition is to accurately translate validation needs into requirements that can be verified. Another way of looking at the difference for an actuator is verification identifies if the actuator as designed and built meets the requirements, while validation identifies if the actuator design meets customer needs. Most test activities are associated with verification because they can be done against specific requirements. Validation activities may utilize results from verification testing with additional work to show how the verified requirements show that the system is validated. Most validation activities are done at the complete system level because that is where an accurate assessment of how well the system meets customer needs can be made. A design verification and validation program should be designed to follow the design process, first making sure technology is stable, then making sure the preliminary design meets a robust set of verification requirements, and finally validating the final design though a series of tests, modeling and analysis to show that it can meet customer expectations through the environmental range it may experience. There are several ways to verify and validate a design, including engineering judgment, comparison, analysis, and test.

7.1 Engineering Judgment

In engineering judgment, the organization is depending on the experience of engineering personnel to evaluate the ability of a design to satisfy a given requirement or condition based on basic engineering principles, prior knowledge, or other experiences.

This means of verifying or validating a requirement is one of the least used because there is no data to back up the assertion that the requirement or condition is met. However, there are cases when it is the only means for verification or validation so that it must be considered.

7.2 Comparison

Comparison utilizes a comparison between the new design and an existing design or product to demonstrate that it meets a specific requirement or condition. This approach involves collecting data on both the new and existing product for the comparison and relies on the similarity between the designs to convince everyone that the design meets the requirement/condition because the existing design meets the requirement/condition. Part of the story may include analysis to show how the designs are similar, and therefore, the comparison is valid. This approach can be particularly useful when empirical data are available, but a correlated analytical technique or testing is not a feasible option.

7.3 Analysis

Analysis is a good option for product verification and validation, particularly when the analytical techniques are well documented and have been correlated with test data. For this verification/validation approach, the design is analyzed against the requirement or condition and the analytical result is evaluated against the pass/fail criteria. Analytical techniques have the advantage of easily performing simulations at different conditions to evaluate how sensitive the design is to each condition. They can also offer an evaluation of the design margin against the requirement and offer a cost effective solution for designs or conditions that are difficult and/or expensive to test.

7.4 Testing

Testing against a particular requirement/condition or set of requirements/conditions is the favorite method of verification and validation. This is because a well-designed test offers observable and conclusive evidence that the requirement or condition has been met. The results can also be used to demonstrate to customers that the design is "real" and operates as advertised.

One of the key components to a successful test program is a good test plan. The test plan is going to evolve as the design evolves and matures. A good test plan includes an objective, an overview of the design to be tested, and specific requirements or conditions being addressed by the test, specifically how the test is addressing these requirements or conditions, pass/fail criteria, test fixtures needed, test equipment needed, test facilities, calibration of test equipment, data collection requirements and equipment, safety, number of people and skills needed to support the test, and schedule showing the relationship between prototype parts and all test resources including test facilities, test equipment, and personnel. Time spent preparing a good test plan pays the program back with clear communication of test goals and accurate coordination of test activities so that expensive test facility time is not wasted, tests do not need to be repeated due to questionable results, and key personnel are available when needed. As the design becomes more refined, the test plans supporting verification activities also become more specific and refined.

Other key components for successful design specific test program are test fixtures. The test fixtures need to accurately replicate the real-world conditions that the test is trying to simulate. Any known limitations to this simulation need to be fully addressed in the test plan. Creating test fixtures to accurately simulate real-world conditions is not a simple task and may involve considerable analysis to demonstrate the fixture performs as required. Test and test fixture safety should also be prominent in the effort. By nature, testing contains surprises and nobody wants these surprises to contribute to test personnel injury or equipment damage. For this reason, participation in tests should be limited to experienced test personnel, design personnel, and any other observers should be located a safe distance from the test. All test fixtures should also be designed with sufficient factors of safety to cover all unknowns, potential failure conditions, and overload conditions. A cross functional review of the test can also uncover any shortcomings that are overlooked by the design and test personnel intimately involved in the test program.

As with the other activities, technology demonstrator testing needs significant involvement from the design organization. The test plan for the technology demonstrator phase needs to challenge the technology being employed in the actuator to make sure that it is stable. Routine design features do not need to be challenged because they have been proved in prior applications. At this stage, testing is generally done utilizing as much generic test equipment and fixtures as possible due to the desire for test data as soon as possible and test interfaces are flexible because the technology is being verified and not the complete design. The desired outcome of this test is a clear understanding of the technology maturity as it applies to the ability of the design to meet the requirements. While it is desired to get confirmation that the technology being employed is mature enough for the actuator to meet the design requirements, it is also valuable to learn that it is not ready or it has limitations. Either way, a clearer path forward for the project comes from the test.

During preliminary design testing, the objective is to verify the design meets all the design requirements. At this point, most testing is still verification testing because it is against the design requirements and not application testing, which would identify requirement deficiencies. The design being tested is expected to be complete and analytically stable. At this stage, the test plan should be specific to the design under test and fixtures should accurately interface with the actuator. The fixture should accurately simulate the conditions to be verified. Data need be gathered to conclusively demonstrate performance against the test conditions. At a minimum verification against the requirements, needs should to be demonstrated. Identifying the limits of the design offers the chance to utilize the actuator under test in a more demanding application without additional testing. When all the preliminary design tests have been completed, the design should be ready to be approved for production.

Ideally, validation takes place during the production readiness phase. Here the design is expected to be stable, having passed all design verification activities. As engineering judgment, comparison, and analysis validation activities can occur prior to availability of a production readiness prototype, the only validation activity left is validation testing on the production readiness prototype. This testing is expected to validate the product, including design, fabrication, and assembly activities. The test plan can focus on customer or user activities to see how the product performs at its assigned task. However, there needs to be a test plan to ensure that the product is not used beyond its intended limits, a data collection system is in place to evaluate actuator performance, the pass/fail criteria is clearly defined. This user testing should be coordinated with the original equipment manufacturer to capture as much of the intended performance and environmental range as possible. With a robust set of requirements and verification plan, these tests should just be confirmation that the design meets the customer's and user's needs.

Chapter 8
Production

Production of a saleable product is the final objective. Ideally moving from the production readiness phase to production should be a straight forward natural process if all the requirements analysis, design, and prototype activities are completed. In reality, this rarely happens because design, manufacturing, and cost improvements can always be incorporated. Design can provide an assessment on the significance of the design improvements and their effect on safety, warranty costs, and company reputation. In the end, it becomes a management function to assess all the potential improvements and decide when the product is ready for production. After production starts, engineering efforts transition from product development to manufacturing support, customer support, and product improvement. These efforts are part of the phase after product development, which utilize the engineering fundamentals captured here.

Bibliography

1. Aerotech Inc. 2010. Linear Motors Application Guide. Aerotech, Inc: Pittsburgh, PA.
2. Beer, F. P., and Johnston Jr., E. R. 1976. Mechanics for Engineers, Dynamics Third Edition. McGraw-Hill Book Company: New York.
3. Boyer, H. E., and Gall, T. L. 1985. Metals Handbook Desk Edition. Metals Park, OH: American Society for Metals.
4. Byars, E. F., and Snyder, R. D. 1975. Engineering Mechanics of Deformable Bodies, Third Edition. Intext Educational Publishers: New York.
5. Commander, Air Force Systems Command. 1969. "MIL-STD-499, System Engineering Management." www.everyspec.com. (Accessed October 2015)
6. Defense Contract Management Agency, Department of Defense. 2012. Earned Value Management System (EVME) Program Analysis Pamphlet (PAP). Defense Contract Management Agency, Department of Defense: Virginia.
7. Leonard, J. 1999. Systems Engineering Fundamentals. Defense Systems Management College Press: Fort Belvoir, VA.
8. U.S. Department of Defense. 2011. MIL-HDBK-881, Work Breakdown Structures for Defense Material Items. U.S. Department of Defense: Washington, DC.
9. U.S. Department of Defense. 1993. MIL-STD-118, Design Guide for Military Applications of Hydraulic Fluids. Department of Defense: Fort Belvoir, VA.
10. Gast Manufacturing Corporation. 1986. Air Motors Handbook. Benton Harbor, MI: Gast Manufacturing Corporation.
11. Harris, C. M., and Crede, C. E. 1976. Shock and Vibration Handbook. McGraw Hill Book Company: New York.
12. Haslam, J. A., Summers, G. R., and Williams, D. 1981. Engineering Instrumentation and Control. Edward Arnold Ltd: London.
13. Hempe, D. W. 2011. Advisory Circular 21-16G. U.S. Department of Transportation, Federal Aviation Administration: Washington, DC.

14. Hulst, S., and Kamstra, L. 2012. Gearmotors: Achieving the Perfect Motor & Gearbox Match. Groschopp, Inc: Sioux City, IA.

15. Hydraulics & Pneumatics. 2015. Hooked on Hydraulics Fluid Cleanliness. Penton Media, Inc: Overland Park, KS.

16. John Deere Service Publications. 1979. Hydraulics, Fundamentals of Service. Deere & Company: Moline, IL.

17. Meriam, J. L. 1975. Statics. John Wiley & Sons, Inc: New York.

18. Monarch, I. A., Goldenson, D. R., and Stoddard, R. W. 2009. An Innovative Requirements Solution: Combining Six Sigma KJ Language Dage Analysis with Automated Content Analysis. SEPG North America 2009: San Jose.

19. MTS Sensors. 2014. Temposonics(R) Magnetostrictive, Linear-Position Sensors. MTS Sensors: Cary, NC.

20. Naval Air Systems Command, Naval Sea Systems Command, Naval Supply Systems Command, Space and Naval Warfare Systems Command, and Marine Corps Systems Command. 2004. Naval Systems Engineering Guide. Department of the Navy: Washington, DC.

21. Nippon Pulse America, Inc. 2012. Basics of Stepper Motors. Nippon Pulse America, Inc: Radford, VA.

22. Nof, S. Y. 1985. Handbook of Industrial Robotics. John Wiley & Sons, Inc: New York.

23. Ogata, K. 1992. System Dynamics. Prentice-Hall, Inc: Englewood Cliffs, NJ.

24. Ohshima, Y., Akiyama, Y., Hata, K., Himemo, H., Masaki, K., Mori, H. et al. 1988. Servo Sensors, Elements and Applications. Intertec Communications Inc: Ventura, CA.

25. Reswick, J. B., and Taft, C. K. 1967. Introduction to Dynamic Systems. Prentice-Hall, Inc: Englewood Cliffs, NJ.

26. Rexroth Worldwide Hydraulics. 1984. Components for Hydraulic Proportional and Servo Systems, Electronics and Accessories. The Rexroth Corporation: Bethlehem, PA.

27. Salmon, C. S., and Johnson, J. E. 1980. Steel Structures Design and Behavior. Harper and Row: New York.

28. Shortley, G., and Williams, D. 1971. Elements of Physics, Fifth Edition. Prentice-Hall, Inc: Englewood Cliffs, NJ.

29. Sperry Vickers. 1967. Mobile Hydraulics Manual. Sperry Rand Corporation: Troy, MI.

30. Sperry Vickers. 1970. Industrial Hydraulics Manual. Sperry Rand Corporation: Troy, MI.

31. Spotts, M. F. 1978. Design of Machine Elements, Fifth Edition. Prentice-Hall, Inc: Englewood Cliffs, NJ.

32. Streeter, V. L., and Wylie, E. B. 1975. Fluid Mechanics. McGraw-Hill Book Company: New York, NY.

33. Swokowski, E. W. 1975. Calculus With Analytic Geometry. Prindle, Weber & Schmidt, Inc: Boston, MA.

34. The Lee Company Technical Center. 2000. Technical Hydraulic Handbook. The Lee Company: Westbrook, CT.

35. Ulrich, K. 2003. KJ Diagrams. The Wharton School, University of Pennsylvania: Philidelphia.

36. Wark, Kenneth. 1977. Thermodynamics. McGraw-Hill Book Company: New York, NY.

Appendix A
Hydraulic Symbols

1. Basic symbols

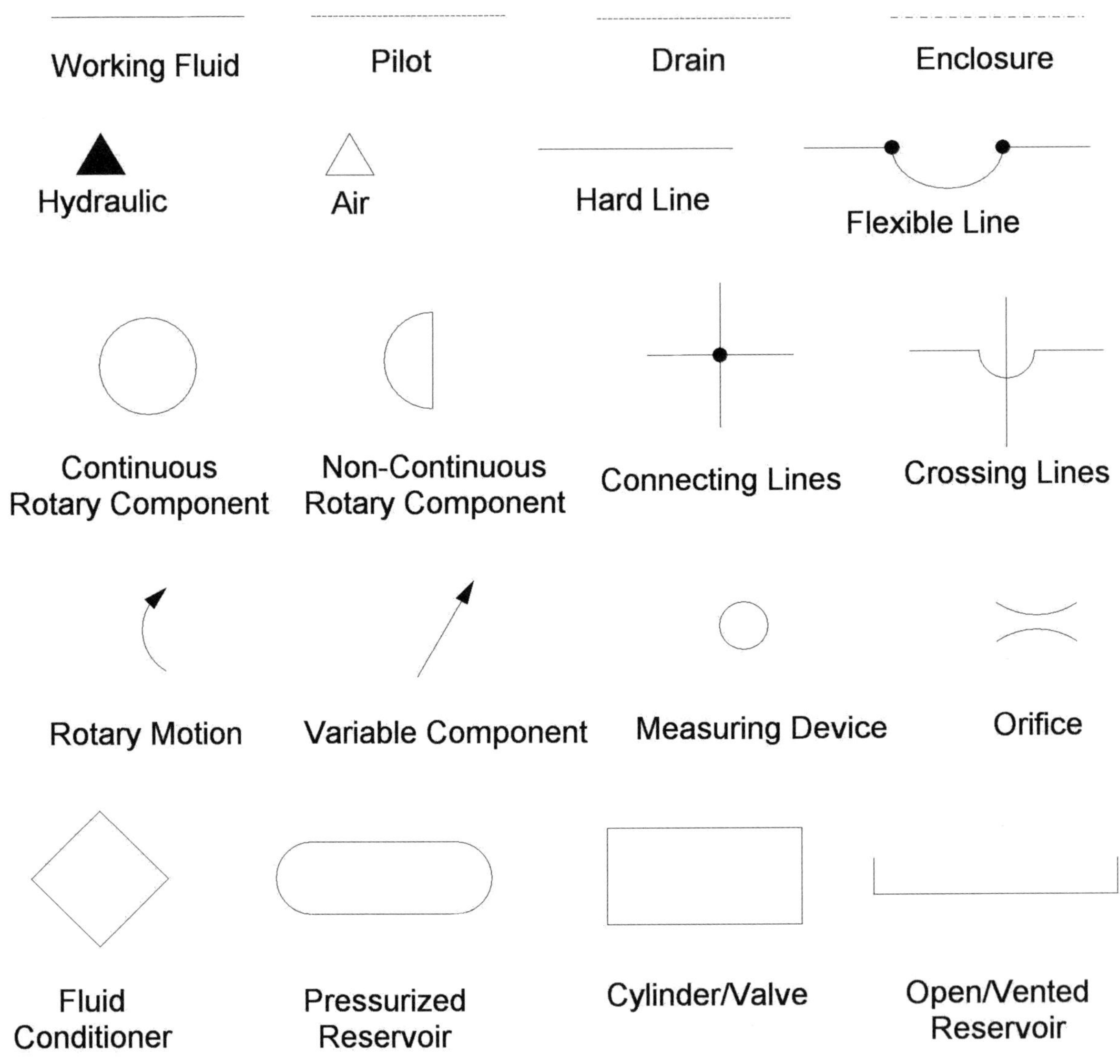

2. Fluid conditioning symbols

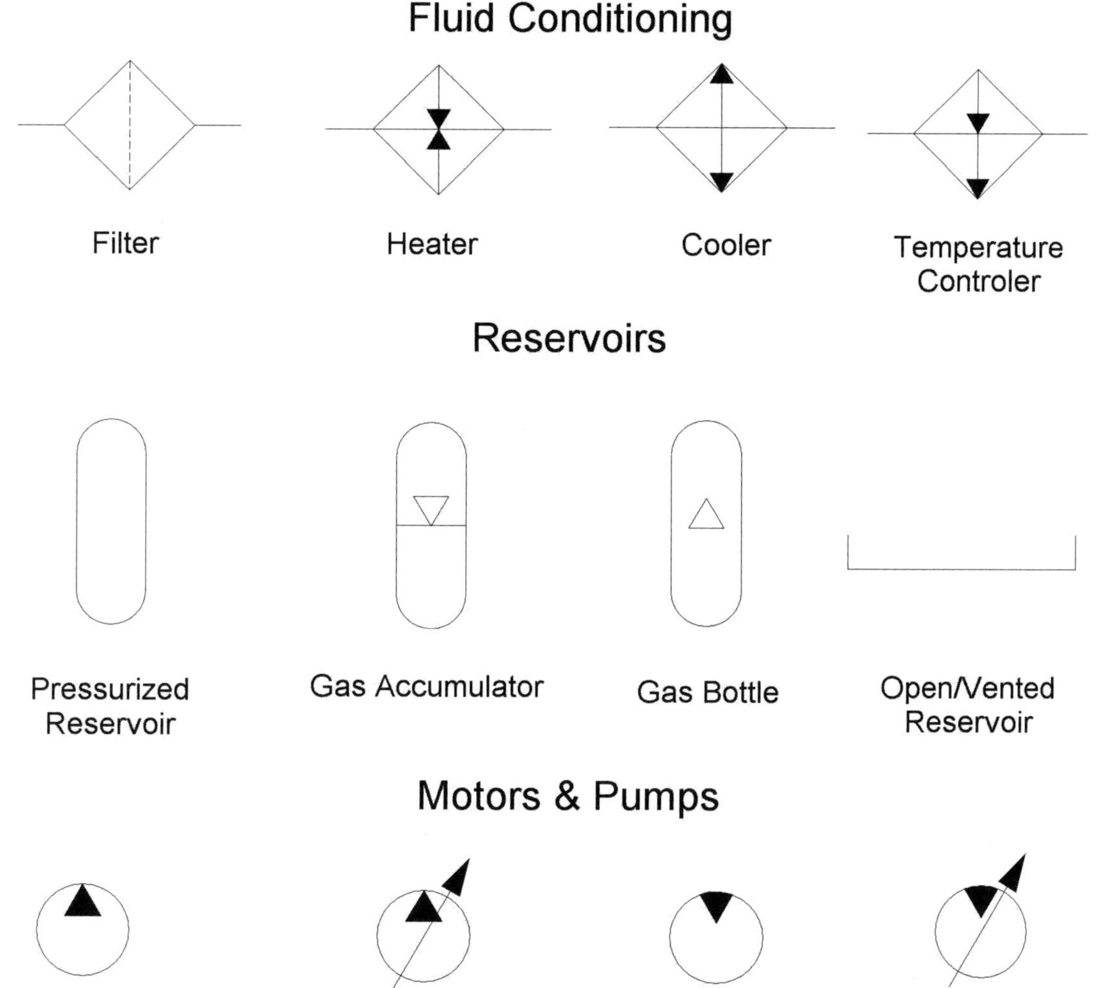

3. Cylinder symbols

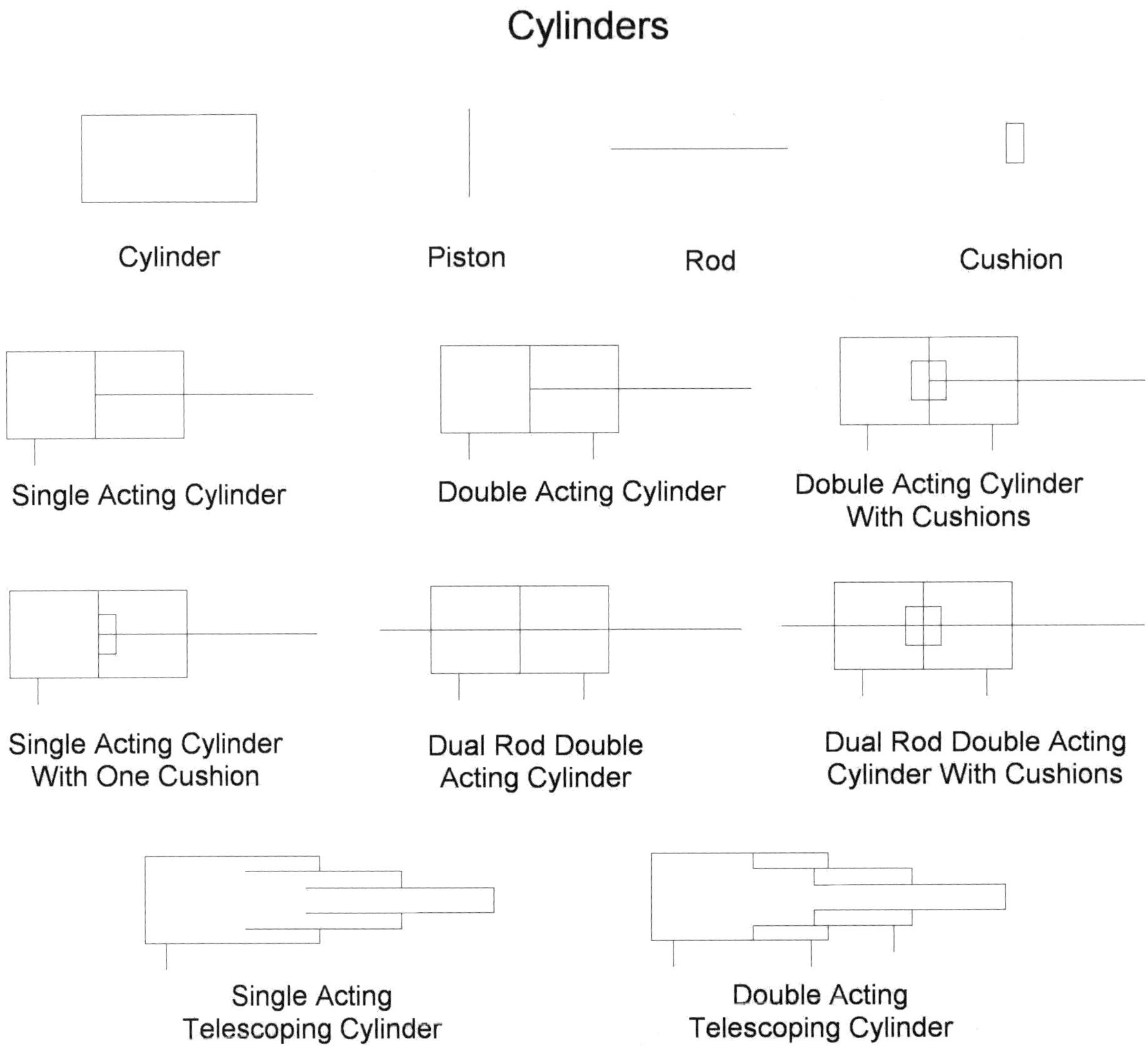

Cylinders

Cylinder

Piston

Rod

Cushion

Single Acting Cylinder

Double Acting Cylinder

Dobule Acting Cylinder With Cushions

Single Acting Cylinder With One Cushion

Dual Rod Double Acting Cylinder

Dual Rod Double Acting Cylinder With Cushions

Single Acting Telescoping Cylinder

Double Acting Telescoping Cylinder

4. Valve symbols

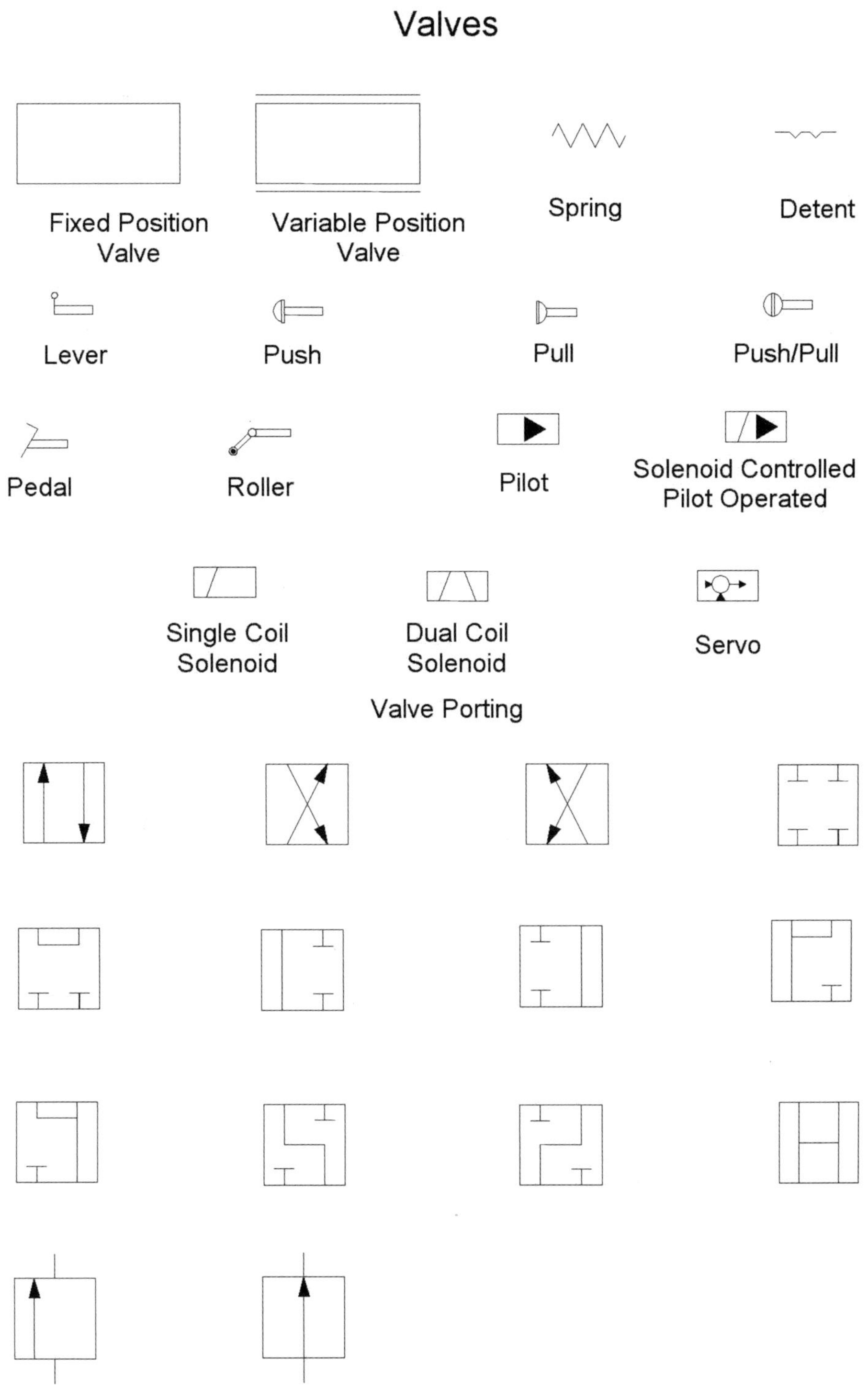

Valves

Fixed Position Valve

Variable Position Valve

Spring

Detent

Lever

Push

Pull

Push/Pull

Pedal

Roller

Pilot

Solenoid Controlled Pilot Operated

Single Coil Solenoid

Dual Coil Solenoid

Servo

Valve Porting

5. Valve type symbols

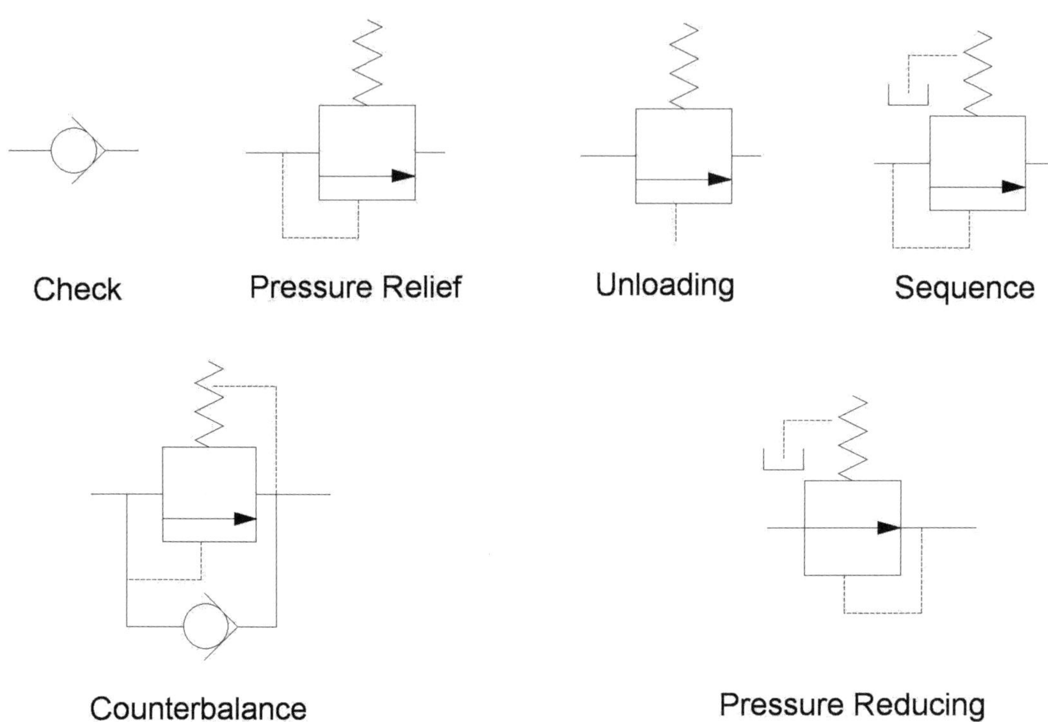

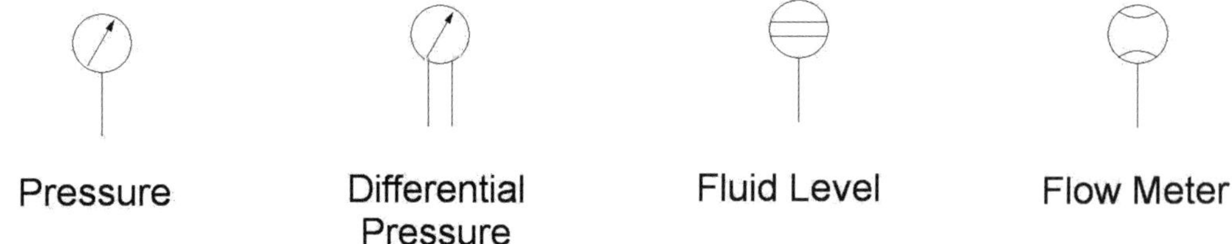

Training Supplement—Problems by Chapter

Chapter 1 Problems

1. What items should be considered before starting the actual design process?
2. What is performance?
3. Which is more important and why, strength or stiffness?
4. What design margin for should be used for a simple indoor actuator when the loads are well known. There is little chance for an overload, analysis is straightforward, and there are no special material requirements, but is there a chance for significant property damage and personnel injury?
5. What environmental conditions should be considered for an actuator mounted on the outside of a hydraulic truck crane?
6. Is minimizing maintenance or making maintenance easy more important?
7. What are the three primary cost components? Which cost component is most important to the owner of a rental agency? Which cost component is most important to a product owner and user?

Chapter 2 Problems

1. What are some techniques to gather voice of customer and translate it into requirements?
2. If a project has completed \$125,000 worth of the scheduled \$135,000 work, what is the SPI?
3. Develop a WBS through the top 3 levels for a high-performance hydraulic cylinder system.
4. List the important considerations hat need to be managed to control project cost.

5. If a project has completed \$125,000 work of work with an actual cost of \$115,000, what is the project CPI?
6. What is your assessment of a project with an SPI and a CPI as calculated in problems 2 and 5?
7. What is the advantage/disadvantage of a bottom-up cost estimate?
8. If a project team estimates that the project is going to cost \$350,000 to complete and the project has already cost \$900,000, what is the EAC?
9. If the baseline plan has an EAC of \$1,150,000, what is the estimated CPI at project completion?
10. What is risk?
11. How can risk be managed?

Chapter 3 Problems

1. How much work is performed on a can of beans when it is moved 10 m up a conveyor inclined at a 30° angle from horizontal?
2. What impulse is required to get a 35-kg projectile moving 750 m/s?
3. How much energy must the brakes absorb to stop a 1600-kg car moving at 100 km/h?
4. If the brakes in problem 3 go through a frictionless mechanism to compress a spring with a spring rate of 12000 millimeter/mm, how far is the spring compressed?
5. How much power does it take to accelerate the car in problem 3 to 100 km/h in 20 s?
6. How far does a car go under a constant acceleration of 6000 m/s^2 for 30 s?
7. If the car in problem 6 has an initial velocity of 30 km/h, how far does it travel after accelerating at 6000 m/s^2 for 20 s?
8. How fast is the car in problem 7 going after the 20 s of acceleration?
9. If a 1600-kg car is initially at rest but experiences a jerk of 0.7 m/s^3 until it is accelerating at 2 m/s^2, how fast is it going after 15 s?
10. If a car initially at rest experiences a jerk of 0.7 m/s^3 until it is accelerating at 2 m/s^2 and then experiences a jerk of -3 m/s^3 until it reaches a constant velocity of 30 m/s, which it maintains for 0.15 s before experiencing another jerk of -3 m/s^3 until it is decelerating at 3 m/s^2 before a jerk of 0.7 m/s^3 brings the velocity to 0, how far does it go?
11. How long does it take a car to complete the move profile described in problem 10?
12. If a final velocity of 30 mm/s is acceptable and a design margin of 4 is desired, what natural frequency does the system need to achieve the final deceleration described in problem 10?
13. A system needs to travel 1000 mm in less than 1 s. It has a design margin of 4 on its natural frequency of 75 rad/s. It has an initial velocity of 25 mm/s. There is an

acceleration of 5700 mm/s^2 before jerk takes it to a maximum velocity of 1800 mm/s. The maximum velocity is maintained jerk takes the system to a deceleration of 6000 mm/s^2. Finally, the system is eased into a final velocity of 50 mm/s by maintaining the design margin on the natural frequency. The time at the initial velocity and the final velocity is 0.05 s. How long does it take for the system to travel the 1000 mm? Show the velocity, acceleration, and position versus time plots.

14. How long does it take to complete the cycle if the natural frequency is 125 rad/s?
15. What is the cycle time if the acceleration in problem 13 is 6000 mm/s^2 and the deceleration is 6500 mm/s^2? What if the natural frequency is changed from 5 to 200 rad/s? What if the natural frequency is changed to 50 rad/s?
16. What motor torque is required to accelerate a load at 2500°/s^2 if the load inertia is 70 kg·m^2, there is a 10:1 gear reduction between the motor and load that has an input inertia of .0005 kg·m^2 and the motor inertia is .0011 kg·m^2?
17. For a design with very high accelerations but low payload, is stiffness or strength more of a design concern? Why?
18. If strength and mass are a concern, what material properties are the most important?
19. If stiffness and mass are a concern, what material properties are the most important?
20. If a bar with a stress of 482 MPa has a strain of .0023, what is the elastic modulus of the material?
21. What is the compressive/tensile spring rate of a bar 50 mm in diameter and 200 mm long? The elastic modulus for the material is 210 GPa.
22. What is the tensile spring rate of a 50 mm × 50 mm steel bar 200 mm long? What is the spring rate if it is made out of aluminum?
23. What is the torsional spring rate of the bar in problem 21 if G = 79 GPa?
24. For three springs in parallel with spring rates of 300, 400, and 500 N/mm, what is the equivalent spring rate for the combination?
25. If the springs in problem 24 are in series, what is the equivalent spring rate?
26. If the springs in problems 24 and 25 represent components in an actuator, what is the equivalent spring rate for the actuator?
27. What is the bending stress in a beam with a bending moment of 25000 N·m and an area moment of inertia of 1.75×10^6 mm^4, 37.5 mm from the neutral axis?
28. For the beam in Figure 3.9, what is the maximum bending moment if L = 1000 mm and P = 50 kN?
29. What is the maximum bending stress if the beam cross section looks like Figure 3.11 with b = 50 mm and h = 75 mm?
30. If the allowable bending stress is 500 MPa, what is the maximum allowable load P?

31. What is the sear stress on the left side of the beam?
32. What is the direct stress under the load *P* if it applied over an area 50 mm × 50 mm?
33. If the beam has a 100-kN tension load *T* applied, what is the resulting direct stress?
34. What are the maximum and minimum principal stresses 5 mm to the left of the applied load?
35. What is the maximum Mises–Hencky stress 5 mm to the left of beam center?
36. For the beam of problem 28, what is the section modulus?
37. What is the plastic modulus?
38. For a maximum stress of 500 MPa, what is the maximum allowable load *P*?
39. What is the combined life for a component that has a stress/life profile as shown in the following table?

Portion of Life (%)	Stress σ (MPa)	Life at σ (Cycles)
35	500	10^6
15	525	$5x10^5$
25	550	10^5
10	600	4×10^4
15	650	10^4

40. What are the most common joint types in actuator designs? What are the advantages of each?
41. What environmental conditions should the designer of an actuator for a combine include in the requirement set?
42. What environmental conditions should the designer of an airplane landing gear actuator include in the requirement set?
43. What environmental conditions should the designer of an actuator for an automated bottle filling machine include in the requirements set?
44. Is the user of a military vehicle more interested in MTBF for critical failures or the overall vehicle MTBF? Why?
45. What ratio of preventive maintenance to repair cost is acceptable to the owner of mining equipment actuators?
46. Is design cost more important to the manufacturer of automobiles or specialized production equipment? Why?
47. What is the relationship between design and prototype cost?
48. What is the relationship between design cost and production cost?
49. What is the relationship between design, production, and decommissioning costs?
50. How important are operating costs?

51. List five risks associated with the design of a high-speed loading system actuator? Which one is the most important to monitor and why?

Chapter 4 Problems

1. What are the advantages of utilizing overlap between operations? What are the disadvantages?
2. What are the nine potential stages of a motion profile?
3. What is the relationship between energy and velocity? Why is this important when designing any actuator?
4. How much work is done moving a 100-kg mass up 5 m? If this is done is 5 s, what is the average theoretical power required?
5. Create a motion profile for moving a 100-kg mass up 5 m in 5 s. What is the maximum power required? How much energy is wasted if there is not energy recovery?

Chapter 5 Problems

5.1 Problems

1. Why is it important to determine the fluid used and flow rates for the entire operation cycle?
2. Create a hydraulic circuit that drives a hydraulic motor and two cylinders, each in both directions.
3. What fluid properties should be considered when selecting a fluid for a high-performance drive? Why?
4. If an actuator utilizing a fluid with a bulk modulus of 1378 MPa (200,000 psi) and a hydraulic cylinder with 1.5 L of fluid, a 250-mm stroke, and a 25-mm bore experiences pressure oscillations of ±5 MPa, how much does the actuator move?
5. A piece of construction equipment has a 75-L/min pump that operates on a 20% duty cycle. What size and type of tank should be selected and why?
6. The hydraulic motor uses 410 cc over the first 2 s. The first cylinder uses 165 cc over the next 1.5 s before retracting during which it uses 130 cc over 1 s. The second cylinder starts moving 1 s after the first one finishes retracting, using 250 cc for a 3-s move before retracting for 3 s where it uses 165 cc. If this cycle repeats every 12 s, requires a minimum pressure of 31 MPa, has a 6 L/min, 35-MPa pump and an accumulator with a 6.9-MPa precharge, what size accumulator is required?
7. How much heat needs to be dissipated from a 21-MPa hydraulic system that uses 410 cc of fluid to move a 7-kg load up 0.6 m with a mechanism that is 75% efficient?
8. What are the most common types of valve designs?
9. Which type of valve is best for metering flow?

5.2 Problems

1. What are the main advantages of a pneumatic system over a hydraulic system?
2. What are the disadvantages of pneumatic systems?
3. How much larger than a hydraulic cylinder does a pneumatic cylinder need to be for the same stiffness?

5.4 Problems

1. What types of loads need to be considered when selecting a motor?
2. What are the two most important motor performance parameters?
3. When specifying an electric motor, what parameters define the maximum operating envelop of the motor?
4. If a 270 V_{DC} electric servomotor is selected for a design, what is the maximum theoretical speed if the back emf constant is .025 V_{rms}/rpm, armature resistance is 6 Ω, and the maximum armature current is 15 A?
5. For the motor in question 4, what torque can be generated at the maximum current if the torque constant K is 0.7 N·m/A?
6. When would you choose a dc brushed motor over a dc brushless motor? Why would you choose a dc brushless motor over a dc brushed motor?
7. If you have a hydraulic system that maintains a minimum operating pressure of 31 kPa (4500 psi), what displacement does a motor need to create 150 N·m (110.6 ft-lb) of torque?
8. If the motor in question 5 is operating at 1500 rpm, what hydraulic power is it consuming?
9. If the motor in question 5 is controlled by a valve that meters the flow in and out of the motor, which adds 80 cm^3 (4.88 in^3) of additional volume to each side of the system and uses a hydraulic fluid with a bulk modulus of 1380 MPa (200,000 psi), what is the drive stiffness of the motor? If the mass moment of inertia is .0054 kg·m^2 (18.46 lb in^2), what is the natural frequency?
10. What would the spring rate and natural frequency of the system in question 7 be if it operates with air instead of hydraulic fluid? Use 28 MPa (4061 psi) for the bulk modulus of air.

5.6 Problems

1. If a solid steel rod with pined ends is 4 m long, what diameter is required to support 5000 N?
2. What is the outside diameter of the rod in question 1, if it is hollow with an inside diameter of 15 mm?
3. If a hydraulic cylinder operates at 31 kPa (4500 psi), what size cylinder is needed to push (extend) a 200-kN load? If the rod has a diameter of 50 mm, what piston

diameter is needed to pull (retract) the load? The rod seal friction is 1.5 kN (337 lbs) and the piston seal friction is 3 kN (674 lbs).

4. Why would one us a closed seal gland? What are the benefits of an open seal gland?
5. If a piston 50 mm in diameter with a 20-mm rod is retracting at 2 m/s with a total mass of 10 kg, design a set of orifice holes to decelerate it to a hitting velocity of 0.1 m/s over a distance of 25 mm. The fluid parameters are as follows: β = 1,379 MPa (200,000 psi), ρ = 886 kg/m^3 (.032 lb/in^3), and the orifice coefficient is Co = 1.052.
6. What is the natural frequency of a hydraulic cylinder with the following characteristics:

   stroke = 500 mm (19.7 in)

   rod diameter = 18 mm (.71 in)

   rod length = 750 mm (29.5 in)

   piston diameter = 50 mm (1.96 in)

   piston length = 75 mm (2.95 in)

   fluid density = 885 kg/m^3 (.032 lb/in^3)

   material = steel and density = 7833 kg/m^3 (.283 lb/in^3)

   Fluid bulk modulus = 1379 MPa (200,000 psi)

   Each side has .0016 m^3 (100in^3) of fluid between the piston and control valve.
7. If the fluid in question 6 has 5% entrained air, what is the natural frequency?
8. What is the natural frequency of the cylinder in question 6 if it is a pneumatic cylinder?
9. What torque is required to provide 8 kN (1800 lbs) of force for a screw with the following:

   pitch diameter = 150 mm (5.91 in)

   lead = 10 mm (.39 in)

   Φ = 30°

   bearing diameter = 200 mm (7.87 in)

   μ_t = .10

   μ_b = .05
10. Is a brake needed to hold a 4-kN (900 lb) load for the screw in question 9?
11. What is the spring rate of the screw in question 9 if it is made from steel, E = 206,843 MPa (30 × 106 psi)? If it has a 200-mm (7.87-in)-long steel nut, what is the spring rate of the nut? What is the spring rate of the screw/nut combination?
12. Estimate the critical speed for this screw/nut combination if it is 2 m (78.7 in) and has fixed ends.

13. Determine the total magnetic flux density for the following solenoid:

    length = 100 mm (3.94 in)

    wire turns = 2000

    current = 5 A

    area = 1600 mm^2 (2.48 in^2)

14. What is the force associated with the solenoid in question 13 for a 5-mm (.20-in) gap?

5.7 Problems

1. If the load driven by a motor through a 5.6:1 gear reduction with an inertia of .5 kg·m^2 at the input shaft is 2,542 k·m^2, what is the load inertia seen by the motor?
2. For a motor with a speed/torque curve as shown below, driving a 50.5 kg·m^2 load with an acceleration of 1,000 rad/s^2, a maximum speed of 25 rpm, and a constant load of 25 N·m, what gear ratio should be used?

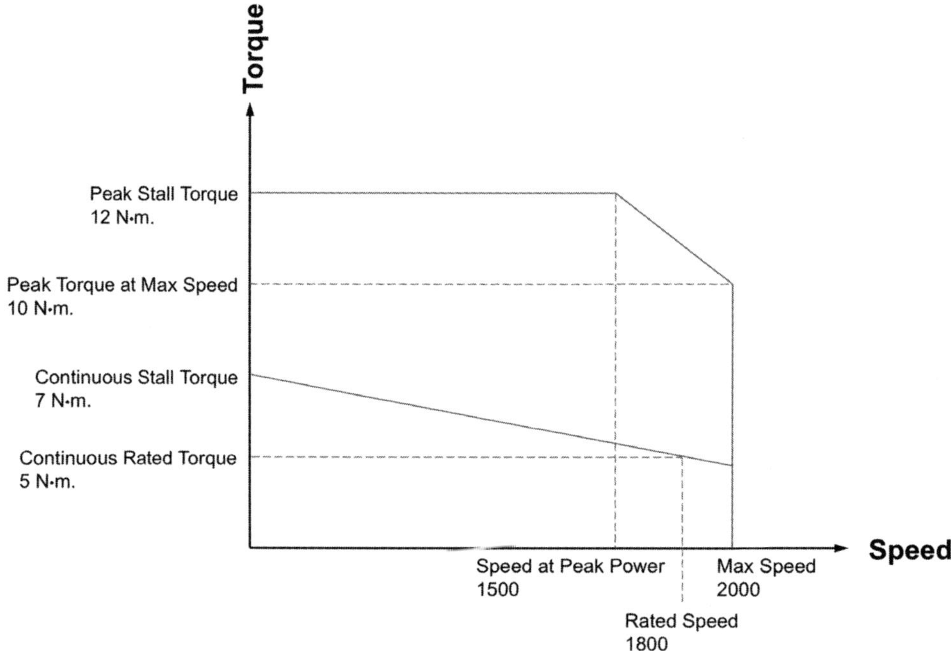

3. What drive spring rate is needed to support smooth deceleration and stopping for the following parameters:

   deceleration = 500 rad/s^2

   velocity = 10 rpm

   J = 25 kg-m^2

   design margin = 4

5.8 Problems

1. List three rotary feedback devices.
2. List three linear feedback devices.
3. List the benefits of locating the feedback sensor at the load. What are the challenges associated with locating the sensor at the load?
4. When would you use a tooth brake?

Chapter 7 Problems

1. When should the manufacturing organization become involved in the design and fabrication processes? Why?
2. When should the purchasing organization become involve in the design and prototype fabrication processes? Why?
3. List reasons to change the design at the production readiness stage.
4. How often should drawings be reviewed for manufacturing and procurement reasons?
5. What are long lead items? How should their design and documentation be timed relative to the rest of the product?
6. What process controls should be implemented as the design and manufacturing process matures?
7. List the major methods for design verification.
8. When is a technology demonstrator prototype needed?
9. Which prototype verification testing phase is the most challenging? Why?
10. What is the difference between verification and validation?

About the Author

The author grew up in rural, Northern Ohio, USA. As a child, his parents could not understand why he dismantled his toys to use their parts to make something different. He moved on to bigger "toys" in his teenage years as he revived an old car and a jeep to drive through the woods and fields of Northern Ohio. He attended The Ohio State University where he received a degree in agricultural engineering. His first engineering position was in the Advanced Products Group of FMC engineering developing cranes and excavators in Cedar Rapids, Iowa. When the downturn in the construction equipment industry developed, he transferred to the Northern Ordnance Division, which later became part of United Defense and BAE Systems in Fridley, Minnesota. There he worked on recoil and ammunition handling system programs such as the Mk 13 missile launcher, Mk 45 gun mount, Advanced Field Artillery System, Crusader self-propelled howitzer, Non-Line-of-Sight Cannon for the Future Combat Systems Program, and Medium Mine Protected Vehicle. Still true to his rural, Midwestern roots, he currently lives on a small farm in Wisconsin and works for Cummins Emissions Solutions in Stoughton, Wisconsin, as the high horsepower product line technical advisor.

The author holds a B.S. degree in engineering from The Ohio State University and a M.S. degree in engineering from the University of Minnesota. He is a registered Professional Engineer in Wisconsin and Minnesota. He currently has seven patents related to his work on automated loading systems and high horsepower emission systems.

Index

Note—Page numbers followed by '*f*' and '*t*' refer to figures and table respectively.